Watersteps Through France

Watersteps Through France

To the Camargue by Canal

With a Foreword by
Libby Purves

Bill and Laurel Cooper
Methuen London

First published in Great Britain 1991
by Methuen London
Michelin House, 81 Fulham Road, London SW3 6RB

A CIP catalogue record for this book is available from the British Library
ISBN 0 413 63740 9

Photoset by Rowland Phototypesetting Ltd
Bury St Edmunds, Suffolk
Printed in Great Britain
by Clays Ltd, St Ives plc

Contents

Illustrations

Foreword

I first met Bill and Laurel Cooper when they were devoting every minute of their life to the barge *Hosanna*, in the wet autumnal surroundings of the Suffolk coast. They were, at the time, putting in a bath. It would have been easy enough to dismiss them as yet another pair of the crazy dreamers whose voyages never get much farther than the mouth of the creek, but from their book on ocean voyaging, *Sell Up and Sail*, I knew better. Bill and Laurel belong to that select band of people who have the determination to turn their dreams into plans, and their plans into actions. Like most long-distance sailors they were a curious mixture of diffidence and confidence, practicality and escapism. Wandering the seas is a pastime as old as humanity and as natural – to some – as breathing; but it takes an immense amount of organisation and determination actually to do it.

They were, like swallows, bound south at the end of that summer for the Mediterranean. Having thrashed across the Bay of Biscay often enough in their adventurous lives, they told us they would take the overland route, climbing the Massif Central through the steps of a waterway system which has long put our own to shame. The idea of crossing such a continent by water, even in the half-built cabin of *Hosanna*, drove me immediately into paroxysms of jealousy from which I have barely recovered.

Still, at least the experience is now shared. Anybody planning the same journey can take advice and experience from their watersteps; anybody mewed up at home, as I still am, with domestic and professional responsibilities at their apogee, can take heart from the fact that the Coopers – with only three good legs between them – left the muddy flats of East Anglia and stormed in such high good-humour through France. What they ate, drank, bumped into and avoided; what birds and animals they saw; what cafés they sat outside drinking *rosé* wine – even what they heard on the radio – is conveyed with sunny, shining good-humour, even to the point of throwing in a few recipes for those of us left at home. I have not yet found *rascasse* or weever-fish for my *soupe aux poissons*, but one can always dream.

Which, after all, is the whole point. The Coopers travel, so that the rest of us can dream. Or, perhaps, plan. Maybe it was unwise of them to give away the fact that *Hosanna* has a guest-room.

Libby Purves

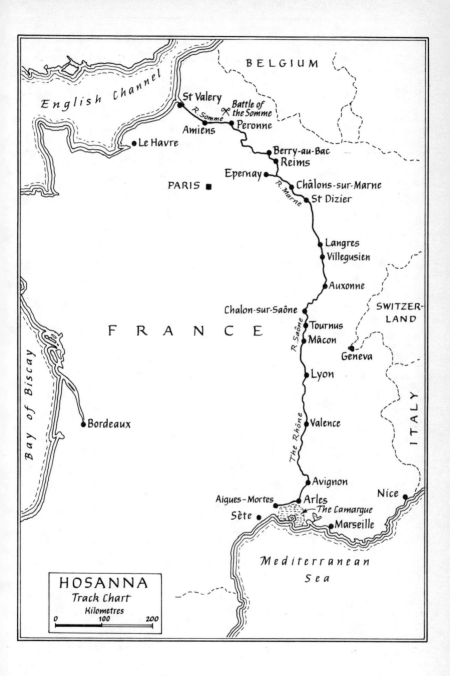

English Channel

BELGIUM

St Valery
R. Somme
Battle of the Somme
Peronne
Amiens
Le Havre
Berry-au-Bac
Reims
Epernay
Châlons-sur-Marne
PARIS
R. Marne
St Dizier
Langres
Villegusien
Auxonne
Chalon-sur-Saône
SWITZER-LAND
R. Saône
Tournus
Mâcon
Geneva
FRANCE
Lyon
Bay of Biscay
Valence
The Rhône
Bordeaux
ITALY
Avignon
Aigues-Mortes
Arles
Nice
Sète
The Camargue
Marseille

Mediterranean Sea

HOSANNA
Track Chart
Kilometres
0   100   200

Preface:
Into France

'Oh lis atiramen de l'aigo blouso, Quand
lou sang nou espilo dins li veno! L'aigo
que ris e cascaio ajouguido Entre li
coudelet . . .
. . . .
'Oh the attraction of clear water
when new blood tingles in the veins!
Water that laughs and purls playful
among the pebbles . . .'*

It would take too long to tell how Laurel, retired art teacher
and ex-magistrate, found herself at a certain age travelling
the world as mate of a sixty-tonne Dutch barge. Let us just
say that thirty-six years of marriage to a master mariner had
something to do with it, but it was not an evolution that
either she or her parents had foreseen.

Enough that she found herself in company with skipper
and husband one summer's day, on a journey that began in
the North Sea and would take us along the French waterways up to the watershed, 320 metres high in the Massif
Centrale, and down again to the Mediterranean Sea,
arriving in time for Christmas.

The barge *Hosanna* is our home. We are travellers-on,
voyagers – what you will to describe a life that depends on
movement through the water, though this story is a

*All chapter headings are from the Provençale *Lou Poèmo dou Rose* by
Frédéric Mistral.

traveller's tale, not a cruising log. Chance encounters delight us, and small dramas occur unsought, to liven the
days. Travelling with an open mind and moderate expec-
tations has much to do with enjoying a journey such as we
now planned.

There are several ways for a boat to travel to the Mediterranean. Why did we not opt for the direct sea-route via the
Bay of Biscay and Gibraltar? Between us we have sailed this
route many times in different ships and boats, and it is not
always rough. An experienced crew and a well-found boat
should be able to cope. But the weather is different in
autumn. It is not the fact that gales are more frequent, nor
that they are stronger and longer lasting, but we contemplated our declining ability to stand rough conditions for a
long time in a boat as yet untried, and caution led us to
consider a less strenuous alternative. Travellers, as the Bard
of Avon said, must be content, and we do not have any
desire to be Great Sailors, only Contented Travellers.

Both the existing routes to the Mediterranean through
inland waters go through France. On the one hand we
could use the Canal du Midi, which runs from Bordeaux in
the Bay of Biscay across to the port of Sète, but to reach
Bordeaux from England one has in any case to cross the
notorious Bay; one is halfway there already and might as
well go on via Gibraltar. Also, in our case *Hosanna* would be
unable to pass under the low arched bridges over the Canal
du Midi. On the other hand we could use the radiating
reticle of canals that carries one southward from the cold
seas off the north coast of France towards the river Saône,
and enter the warm Mediterranean through the Rhône
delta. This was the route we chose.

We like France, and we had never travelled its waterways, nor climbed its watersteps. On the canals shipwreck
and total disaster are unlikely, though even this route is not
without its discomforts and dangers. Although the navig-

able rivers of France have mostly been tamed and canalised now, the concept of taming can be somewhat theoretical. A so-called 'tamed' river can still provide some very exciting navigation when God wills. If you think that on the water-ways the kettle will always stay put and the plates will not slide, you could be surprised, especially on the Rhône south of Lyon, and in winter. The bigger locks can be a rough ride, too.

Though we knew that the inland waters might not be a bed of waterlilies, we did not expect them to be quite so unlike the descriptions in the guide books, all illustrated with summer sunshine, tanned girls in shorts, and charm-ing riverside restaurants with shady terraces where you can eat a lazy lunch or drink a cooling *pastis* under the leafy trees. It is all true, but not in winter. The middle of France can be bitterly cold, canals can freeze solid, the tourist restaurants in the Michelin Guide are closed, and the wily French are all inside by big log fires, swigging last year's sunshine in great goblets of local wine.

Why do we indulge ourselves in this occasionally uncom-fortable pastime of living and voyaging on the water at all? Because we like it. There are too many good things about it to list them all, and overcoming the few adverse aspects adds a spice of achievement to a life that can be too bland.

We do it because we like boats, like living on the water, and like the freedom, the peace, the unexpected and the comradeship. Our combined past experience over many years shows that only on one day out of three hundred, on average, are we as uncomfortable as any commuter going to work in the morning, or any housewife shopping in the rain.

1

Hosanna starts gently

'Eh! Coumo vai lou viage? A la coustumo!'

. . . .

'How goes the voyage? As usual!'

Among the craft at the landfall buoy off St-Valery-sur-Somme, one warm August afternoon, was the Dutch barge *Hosanna* in which we live.

We had arrived after a calm but foggy overnight passage from Great Yarmouth at this little port on the northern coast of France, named after a disciple of St Columba who came from Cornwall to convert the French and built an abbey in AD 610 at the place now called after him. His name was Walric, the French called him Valry, and that is why the 'e' in St-Valery is not pronounced.

Bill, the skipper, had chosen St-Valery because few people enter the French canal system by the river Somme. He hates crowds, even small ones. Whenever two or three are gathered together in anyone's name he goes somewhere else.

St-Valery is an awkward port to enter. No one has ever offered a convincing explanation for the fact that tides on the north coast of France have a much greater range than those opposite them on the south coast of England. That is to say that French high tides are higher, and low tides are lower, than those only a comparatively short distance away across the Channel. We think the French introduction of the metric system may have confused the Almighty; He probably looked up the figure eight in His Heavenly Tide

Tables and gave the French eight metres against a paltry eight feet for the English.

Whatever the cause, there are few northen French ports that are accessible by boat for twenty-four hours in twenty-four. Usually one is obliged to wait offshore for sufficient depth of water to appear, then hurry like mad to get into a safe berth before the plug is pulled and the water disappears – and quite often it does just that. Many ports like St-Valery-sur-Somme dry out completely at low water, and what appears at high tide to be a bustling and lively scene of nautical activity turns into an acre or two of mud or sand, with stranded boats scattered here and there where chance has left them.

Out to sea off St-Valery and all such drying ports (and they are always the more attractive ones because *commerce* goes to the ever-open doors of a deep-water port), each rising tide finds a small flotilla of miscellaneous craft moaning at the bar, waiting with varying degrees of patience for enough water to float them over.

We had arrived a bit early to be sure of being there at the right time. Idling with us at the start of the serpentine channel into the Baie de la Somme were three or four fishing boats, a half dozen or so small French yachts and a couple of gallant Brits, all of whose occupants were engaged in a strange war dance, beating about them with newspapers, frying pans or aerosols of 'Fly Die'.

This exhausting activity on a warm summer afternoon was caused by an extremely dense offshore waft of mixed insects, apparently originating in Berck-Plage, a seaside town a little to the north. Some (the butterflies) flapped about looking pretty, but most of them bit, stung, tickled, scratched or crawled under our clothes into places we would have preferred they hadn't. We swept up a panful of them when the cloud had passed: large and small, chitinous or squashy, winged or fanged, together with a rustly pile of detached wings and legs. So often, when you think you might be bored, life provides you with a cabaret.

Suddenly, the first fishing boat burped its engine into life and headed confidently into the channel, hotly pursued by the company. It had passed some two or three of the marker buoys when it went suddenly aground, and the fleet came to a confused halt, closing up behind it like greyhounds when the electric hare breaks down. We milled about, very close together, no one wanting to push ahead, lacking confidence now that our assumed leader had turned out not only to have feet of clay, but to be firmly stuck in it.

Ten minutes later another fishing boat made a rush for home, and suddenly we were all after him. This time there was enough water and the fleet strung out. A few boats seemed bemused when the channel bifurcated, and two or three craft circled, as if unable to believe that there were two channels and that the lead boat was not going their way. Then, with the Le Crotoy contingent moving off to the eastward, the rest of us continued up towards St-Valery, separating a little, for as the water deepened boats felt able to use their maximum power. As we drew closer to the port we met the outward-bounders; friends waved and shouted genial abuse amid boisterous laughter, while the British yachts' crews waved genteelly, not wishing to make a scene.

We had chosen to arrive at the period of neaps, when the high tides are at their most modest. The St-Valrians, who have been watching tides for a long time, have built their quays so that the highest of high tides has to struggle to get to the top; about eight metres. As we berthed at the fishing quay (we were too big to go into the yacht harbour further up) we were effectively at the foot of a high prison wall.

Fortunately, unlike a prison, ladders had been built into the wall, and after making fast we shinned up and said hello to the statue of Bill's namesake, the Conqueror. It was from St-Valery that William of Normandy sailed on his voyage to beat Harold and become King of England, and the St-Valrians are proud of it.

It was getting late and the sun was setting; there were desperate thirsts among *Hosanna*'s people, and rumbling tums to fill (we had been joined for the sea crossing by our son Ben and friend Alan). The four of us spruced ourselves up a bit, fed the ship's cat and then, wiping off the seaweed we had collected from the ladder, were welcomed into the restaurant of the Hôtel du Port. We ate and ate and ate. We even drank a little, too, blissfully discussing the previous night's crossing; we had all stood our watches, but Bill in particular had had very little sleep because of fog. We had had a very hard few months preparing to leave England, and our thirty-six-hour voyage down the North Sea was not without its hazards; much time was spent dodging dragons in thick fog, as the cross-Channel ferries appeared to us in the Straits of Dover.

Now we were abroad and travelling again, and it felt good. At the moment, however, we could not proceed. Because a sea-lock gate had deteriorated, and one of the pairs was no longer safe, we had to wait about a week at the fishing quay for springs, the period when the high tides are at their highest, so that the levels on each side of the lock gates would be nearly equal, thus minimising the pressure on the faulty gate. This problem of lack of maintenance on the French canals is one that crops up again and again; some of the waterways are clearly moribund. The St-Valery sea lock was due for repair later in the year.

Since we had to wait, we decided that we had earned a holiday, if only for a week, and settled down to enjoy St-Valery and the amenities of the area.

Being accessible by boat for only a couple of hours near high water gives a gentle, semi-diurnal rhythm to the port. Twice each day people gather on the quay, where the little harbour opens off the wide expanses of the Baie de la Somme.

First the departing craft start their engines, with a cough and a roar and puff-clouds of blue-grey smoke. Ropes are cast off, there are farewell shouts to friends on shore – '*Au 'voir! A bientôt!*' – and then the craft disappear evenly down the narrow channel, their arrow-head bow-waves cannoning off the banks to cross and recross, weaving a net with diamonds in it.

Then after a time the incomers appear, the roar of diesels increases, and round the corner the first fishing boats – the *sauterelliers*, so called after the little grey shrimps they catch – come home with cries of '*Salut, les gars!*' and '*Ça va?*'. Then come the yachtsmen in their fancy yellow oilies, and one can hear fluting tones from a red-ensigned boat saying: 'Jeremy, where did you put Timothy's passport?'

The fishing boats make fast, their catch is lifted on to the high quay and trundled into the covered fish market a few yards away in Rue de la Ferté. Then all goes quiet until the next tide.

As the waters recede the boats sink out of sight like department-store lifts, and nearly as fast. The anglers re-appear. Families with small children promenade, the little ones in bright holiday shorts being dragged back from the edge as they make foreseen but dramatic darts towards – not death by drowning, but a long drop into muddy sand.

The elderly, the idle, the curious and the tired sit on the quayside benches with their dogs and baskets of shopping; passing the time of day, eating ice cream or hamburgers and hot dogs from 'le fast food' stall (that France should come to this!).

It is surprising that there are not many British tourists at St-Valery. We found the inhabitants most hospitable and friendly, and the more traditional catering establishments very good value after the inflated prices of English restaurants of much lower quality.

We returned to the Hôtel du Port again and again to taste, among many other good dishes, the *ficelles,* a Picardian pancake that usually contains ham and béchamel sauce, though the Hôtel du Port had a mouthwatering seafood variation that arrived with cheese bubbling on top. (We have collected recipes for some of the dishes that we enjoyed on the journey, and you will find them at the end of the book). We were soothed also by the comforting ministrations of *la patronne*, Mme Kappel, a cheerful lady of impressive dimensions and an apparently inexhaustible wardrobe of vertically striped dresses, none of which prevented her from looking like a galleon in full sail as she marshalled her forces. These consisted of Yolande of the brilliant smile (who gave our England-bound crew a lift to the station at Noyelles), Patrick the industrious and François-Xavier the joker, who, when asked for a second spoon so that Bill could share Laurel's chocolate mousse, brought one the size of a garden spade.

To prepare for the next stage of the journey we bought a map of the Canal de la Somme at the bookshop. The proprietor apologised, it was the last one he had, and, what a desolation, the cover was not so clean. That makes nothing, Laurel said, it will all be the same when our cat has marched upon it. Ah! he said, you must be from the *péniche* (barge) with the little tiger cat. Laurel had some trouble with the French for tortoiseshell, which is not what you might think when applied to cats, and agreed.

We had acquired our ship's cat the previous autumn, when we were still living ashore in Great Yarmouth while the steelwork was done on *Hosanna*. One blustery November day, after dusk, the door bell rang. It was Chris and Zerda, the new owners of our first boat *Fare Well*, on their powerful motorbike. From inside his leather jacket Chris brought forth a small mewing tortoiseshell kitten and a tin of catfood – a cat-kit from *Fare Well*, as they put it. We had then been catless since the death of Nelson, who had departed this life two years previously in Italy after ten years of sailing with us. We had grieved deeply and buried her with honours in the mouth of the Tiber river. The new kitten's colouring reminded Bill of a half of two's, as mild and bitter is called in our part of the world. She was also our second ship's cat, so Twosie she became, submitting goodhumouredly to going everywhere with us in a wicker basket, usually to work on the barge, which she loved. She grew up accustomed to the bustle and noise of the boatyard, to the crash of falling steel and the rip and spit of welding, and learned to catch mice in the marshes.

Now here she was in St-Valery, a charming small town. Many houses are half-timbered and some have wrought-iron balconies poised on their brick bosoms like black lace, pinned in place with pots of geraniums. We sniffed appreciatively the salt wind off the sandflats, a wind that once ruffled the papers of Anatole France as he sat at his open window overlooking the Baie de la Somme.

The winding channel we had followed on our arrival meandered across the very extensive sand banks of the bay, which were uncovered at low water. Behind the square miles of crustacean-rich sand there were other banks that had hitched themselves a little higher, and now only occasionally tossed a veil of seawater over their shoulders. These had embroidered themselves with sea-grass and marram, as well as collecting a narrow fringe of the slowly decaying products of the packaging industry.

Sheep grazed and dozed. The famous *pré salé*, or salt-marsh lamb, comes from these salt flats. Families with nets went cockling. Wildfowl were there in abundance, thanks to the nearby nature reserve, and where depressions allowed the water to lie the French had built hides, or *huttes*. At one time these served only to provide a means of maximising the slaughter of the birds, for *la chasse* to the French is less a question of hunting than an exercise in massacre. The *huttes* are neatly disguised like gun emplacements under swathes of turf. Happily there are Green people in France too these days, and with them are bird-lovers and twitchers, botanists and naturalists; part of this area is today a reserve, where hunting and shooting are forbidden.

We were much taken with the brochure of this reserve, the Marquenterre Bird Park, which is a crossroads for migrating birds. We quote the English version:

> Visit the bird's land: in the North of Somme . . . an enclave gained by man on the sea. This passage place, which is also a privileged place for sojourn, enables to observe numerous birds species . . . for neophytes or person in the know. The small circuit is an initiation walk. It begins with the shock promoted at the 'view point' with the contemplation of the landscape immensity, the park, the polders, the bay of Somme and its spectacular lighting.
>
> The visitor then runs along ponds and pools where numerous birds are living, and thus attracting their tramping congeners.
>
> The nature, always renewed, lives to the seasons' rhythm . . . plunged into an intimacy with birds, a fascinating scenery, appreciated by artists, photographers, scientits [sic] and nature lovers.

This is the very stuff of language; there is a richness here that is lacking in translations that are too perfect. 'Scientits' has gone straight into our vocabulary to describe ornithologists of the battier sort, and we ourselves, with our wandering life, are tramping congeners, though whether we attract our fellow passagemakers as the birds are said to do is not for us to say.

We experienced further 'shocks promoted with the contemplation of the landscape immensity' when we took the old steam train, Le Tortillard, its glands and stuffing boxes leaking feathers of gently hissing steam, whose ancient wooden carriages smelt like the trains of our childhood, redolent of smuts and varnish, and rode right round the bay on a long, smoky trip across the marshes and occasionally into the cornfields. We had a wonderful view of the bay with its salt flats, and glimpses of a timeless countryside of farm tracks, hedgerows, copses and golden wheatfields; rough lumpy pastures for the rough lumpy cows of the region; willows with their backs to the wind like blown skirts; and a salty pond left by the receding tide, throwing into the air a flight of assorted ducks, like a handful of confetti tossed in the wind. Over it all raced an enormous sky of blue, patched by busy clouds.

Just before the station at Noyelles appeared a pernickety kitchen garden with coloured stripes of tilled earth, sharp green lettuce, purple beet, the straight green stems of spring onions, feathery carrot tops and blue cabbages – as neat as the stripes on Mme Kappel's dresses.

While we sat down for a cup of coffee at the terminus (graced by the ideal holiday clock – one with no hands) a wizened and whiskered crone with nutbrown shins, pushing her possessions in a wicker trolley, sang to us in a voice of haunting beauty, not for gain but for the pure pleasure of singing. Our English tones caused her to break into 'Roses of Picardy', then 'Tipperary', and with a heart-catching jog of memory we recalled that we were close to the battlefields of the Somme, and that our river journey would take us through them.

We returned to *Hosanna*, our floating and peripatetic home. She is larger than the narrowboats which ply the English canals, but smaller than the typical French canal barge or *péniche*. Whereas the narrowboats are about 21 m. by 2 m. 10, and the standard French *péniche* is 38 m. by 5 m., *Hosanna* measures about 26 m. by 4 m. 50. For those interested in such things, there is a plan in the appendix.

Being a Dutch-built vessel capable of serving the offshore islands and crossing the wide expanse of the Ijsselmeer (once called the Zuider Zee), which can be very rough in bad weather, she is built with sea-kindly lines and doesn't much resemble the typical blunt-nosed, square-backsided barge of France or Belgium (known to the Dutch, who have an eye for a boat's lines, as a 'spitz', though this is not as contemptuous as it sounds).

We had bought her while she was still in trade, and had sailed her across the North Sea to Great Yarmouth one cold and choppy day in February, where we had found one of the good old marshland boatyards of Norfolk, Bure Marine, to do the major structural work of her conversion as a floating and mobile home, which they accomplished very well.

We have lived in a boat since 1976. For the first ten years we cruised the world in our thirty-tonne sailing ketch *Fare Well*, which had gradually become a little hard to manage as we got older, so we had decided to change to a more sedate style of nautical living before anything went wrong.

Hosanna at sixty tonnes is bigger than *Fare Well*, but she has a good engine and other mechanical services that ought to make the travelling life easier for a pair of ageing crocks. No more heroics now; life is to be *doucement*, *doucement* – a gentle journey.

Conversions take time, and though the steelwork was all done the accommodation was still unfinished at Great Yarmouth when we lost our nerve at the prospect of another winter on the windswept marshes of Norfolk. Suddenly seeing the year shorten ahead of us, we had arranged for the boatyard to complete the major part of the basic conversion, wheelbarrowed all our gear and the spare parts needed to finish the job into the boat, and nailed up most of the doorways. Then, blessed with a window of fair calm weather, the four of us sailed down the North Sea and the Channel, non-stop for St-Valery.

The result was that our unfinished living quarters were somewhat spartan, to say the least, at the start of the voyage. They consisted of a bed, a sink, a marine toilet, a wood-burning stove, and very few dividing walls. This state of affairs gradually improved, and even became weather-tight with the onset of autumn; we worked as we travelled.

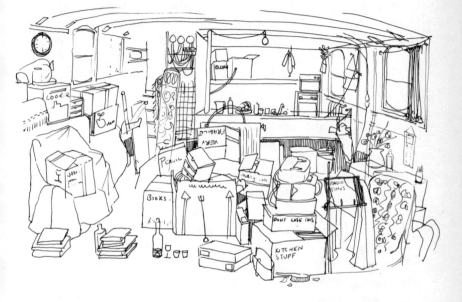

Starting from the forward end of the boat, we have a big store, separated from the rest of the ship by a watertight transverse wall, which is known in the trade as a collision bulkhead; it is there as a safety measure in case we hit anything head-on.

After that there is a bathroom, and then our bedroom, the full width of the ship and about three metres long. We then have a living-room, more often referred to as the saloon, which is six metres long and contains the wood-burning stove. Then comes the galley, which is about two metres square, fitted with an electric oven, a hob with two electric and two gas rings, a microwave oven and a fridge. When finished, it will have thirty-one drawers under the counter tops, for drawers are much easier to use than cupboards in a boat, even though they are a devil to make and hang. After the galley there is a guest-cabin, with its own bathroom.

Underneath the floorboards we have water and fuel tanks, about 5000 litres of each, and a deep-freeze. These are contained in a big storage space which we call the cellar, as neither of us is much given to overdoing the nautical jargon.

After the guest-room is the engine-room, with a big Volvo diesel and a smaller Perkins as a back-up. There, too, is the generator that powers the ovens, fridge and so on – all the noisy, dirty stuff that makes everything work.

Finally, after the engine-room comes the original crew's cabin, called in Dutch the *roef*. This had been the home of a family of four in a total space of some three and a half metres by four. Apart from a tiny living-room they had had two bunk spaces inside cupboards, a miniature galley and an enormous diesel stove, but no lavatory or bath. This is where we had lived while bringing the boat from Holland – in comparison with which our present circumstances, however lacking in such niceties as cupboards, drawers and walls, seemed extremely plush.

Sad to say, the original woodwork in the *roef* had been damaged by woodworm and the carelessness of the day-workers who were running the boat in her last few years in trade, and we had not felt able to restore it. One day it will become a studio and spare guest-room, but for now it was yet another store. The boat was an odd mixture of luxury and squalor. As there were no cupboards yet, our possessions were still in cardboard boxes; and as the interior was not yet lined with panelling, people had been known to peer at the exposed insulation, as lumpy as a Christmas grotto, and remark on the originality of the wallpaper.

All in all we had about 90 square metres, (or about 900

square feet) of living space, which is about the same as a small bungalow.*

On deck we had a wheelhouse, with double doors at the back opening on to a covered verandah. We cheerfully do without a garden, managing with some potted plants and the whole watery outdoors beyond our bulwarks, the rustle of reeds and the flicker of reflected light.

Hosanna was still waiting for the right tides to continue her journey. For every foot the high tide was higher, the low tide was lower; and then *Hosanna* ended on the muddy bottom by the harbour wall. The bottom was not level, so for a few hours each night we slept in a bed that sloped sideways, which is better than no bed at all but not much.

We sometimes breakfasted on the slant, which was familiar from seagoing days. You don't fill the coffee cups too full, nor eat any cereal that rolls out of the bowl before you can stick it down with treacle, and you fry on the downhill side of the pan where the fat has collected.

Hosanna at these times also slid downhill, away from the quay wall, which made climbing up the ladder to go ashore a feat of considerable agility, necessitating hoisting or lowering Laurel's walking stick and shopping bag on a piece of cod line. (Laurel has carried a dislocated hip around with her since birth; she is attached to it, you could say. Having only one left leg, she regards it with a mixture of tolerance and exasperation, trying by willpower to ensure that it knows its place and stays there. Sometimes she succeeds.) Twosie, who had spent the entire sea crossing sensibly asleep down below, was cautious about exploring, and seemed content to sit on the cabin roof and smile at the admiring crowd.

Then the tides came right, and we went through the lock and into the ship canal – a grand description, and in need of qualification, for the only ships that can use it are the

* There is a plan of *Hosanna* on p. 227.

smallest of coasters, which have now virtually ceased to exist, so the commercial traffic has disappeared. We moored at a granary quay and engaged a marine engineer to check over a few problems that had worried us on our maiden voyage down channel. We were in no hurry.

Twosie ventured ashore for the first time, a biddable but curious half-grown cat. Although the roadside track was used only by the boatyard and the granary, we worried a little about her road sense. We watched her; she seemed to be crossing the track carefully, and after her long confinement aboard, first for the sea crossing and then the steep harbour wall on the sea side of the lock, was clearly delighted with her surroundings. A cat is a cat, and one should not confine them without cause.

Across the track there was an old orchard behind two brick sheds, the glass long gone from the windows; there were no roofs except for a half-attached strip of corrugated iron that creaked and flapped when the wind blew. This piece of wild ground, dotted here and there with writhen fruit trees, was surrounded by blackthorn, elder and brambles, and contained a fieldful of wild flowers.

We had watched Twosie leap through the glassless windows into Paradise and remain there for hours; and one warm afternoon Laurel scrambled over the sill and followed her. Twosie seemed pleased with her presence and showed off shamelessly, tearing round a series of rabbit tracks in the long grasses, bounding seven cats high over the wild parsley and fennel stalks, nosing blissfully through the bedstraw, occasionally pausing to pose on a clover cushion; a tortoiseshell cat surrounded by a ruff of fool's parsley like the Virgin Queen in a lace collar. Finally she threw herself on to the warm stone sill of one of the sheds and lay there, a cat and a half long, and Laurel left her, high above the vetches and the yellow, white and green summer sunshades of chervil, burnet and caraway, relaxed but watchful.

Back at the barge we rigged a derrick and lifted ashore our

ancient little Dutch Fiat 600 car, which had crossed the sea on our deck, and explored the countryside. We had rather longer to do this than we expected, because the sun continued to shine and our marine engineer disappeared for a few days to help his father with the grain harvest. We did some painting of the ship's upperworks as well, a job that seems almost continuous in fine weather.

We had watched people setting out long trestle tables in the lofty mediaeval Salt Barn on the quayside for a Kermesse des Poissons, or fish feast, in aid of the Fishermen's Wives Association. We supported it, as fishermen's wives of any breed have a very heavy cross to bear.

The poor ladies were having to cope with faulty gas burners and the feast was a little slow to arrive, but the mussels, mackerel and chips when they came were delicious, borne by the Wives in local costume: dark gathered skirts, crisp aprons, and the pleated fan-like lace bonnet of the Boullonnais district.

Most of the company were good-humoured, and the wives handled those who were not with great firmness. We saw one doughty lady deal with an importunate group calling for their order with some asperity, by plonking bread and wine before them and shouting *'Mange!'* as one would to a demanding child.

Chatting to our neighbours, we found out why the French fishermen receive such support from their government, while ours have none. It is not that the French are better organised than the British – given the life they lead it is virtually impossible to get a fishermen's union together – but in France it is the wives who are the terror of the politicians. This is an apt use for the expression 'a monstrous regiment of women', who pester, abuse and even box the ears of their representatives when they do not deliver the desired results. The French have cultivated a healthy disrespect for politicians and *fonctionnaires*; these creatures have no decorative purpose, they exist only to be used, preferably by the redoubtable ladies of the Fishermen's Wives Association of the Baie de la Somme.

One day as we lay at our mooring we were hailed by four small motor boats from the Royal Naval College at Dartmouth. We sat on the grass with a beer or two, chatted, and took some of them to the *supermarché* for supplies. As they left the following day in a light drizzle one boat departed upstream with her crew in the shelter of a big multi-coloured golf umbrella, a charming informality that would have had the old-fashioned, beetle-browed admirals of Bill's days chewing the eyepieces off their telescopes in dismay.

The days passed. We did some more exterior painting, and a bit of varnishing here and there, assisted by Twosie, who stepped in it, helpfully.

After several days of reminding each other that we had crossed the Channel with the idea of reaching warmer climes for the winter, we felt it was time to make a start on the journey. We took on water, and stores. With our three masts now lowered, we committed ourselves to the canals of France and the inland route to the Mediterranean.

Our intention at the start of the voyage had been to seek a comfortable mooring in the south of France to enable us to finish the conversion in slow time and warmer weather, and this was the aim of our present journey.

No more, we hoped, would we be trying to paint the outside of the boat in the steady August rain of England (the previous summer had been exceptionally poor). We would exchange the boatyard noise of welding-spit and crashing steel for birdsong and watersong; mud and rust for the healthy country muck of moist riverside earth. And instead of the smell of hot metal and diesel that composed the boatyard's aura, our grateful noses would be soothed by meadowsweet and mown hay.

We were travelling hopefully again, following the swallows down to the sun.

2

September Somme

'E, las! d'aqueu varai tout ço que resto Es
lou traçan e la rousigadouro Que la maio
a cava contro li peiro.'
. . . .
'Alas! all that remains is the etched
trace, the groove that the cable has
worn in the stone.'

It was early on a sunny Sunday morning when *Hosanna* left
St-Valery to begin the journey down the waterways of
France. The first bridge a hundred metres away should
have been a lifting bridge, but like so much else on the
canals it was *en panne*, its electric motor having been ruined
in a recent thunderstorm. On advice from the sealock-
keeper we had to seize our chance and make our way under
the sidespan at very low water, though not so low that we
ran aground, a tricky decision involving considerable
judgement and a certain amount of trust between the two
people on board.

After nearly forty years of mutual boating in craft of all
sizes, and fifteen years of living aboard, you could say that
we understand each other.

Bill had fixed a guide pole upright in the bows.

'Twang pole,' he said succinctly. If that would go under,
the rest of the boat could follow.

Although Laurel did not doubt that the skipper had
measured the pole correctly, it was a panicky moment as we
slowly approached the bridge, waiting to see if our pole
would 'twang', and prepared to reverse rapidly if it did.

We cleared the bridge with a few inches to spare, and were on our way up the river Somme. On our way – past alder hedges, banks of willowherb and huge parasols of hogweed, where coots chirrupped and waddled in the mud at the water's edge. On our right was the towpath, lined with poplars and punctuated by Sunday fishermen.

At Petit Port the lift bridge was opened for us, and the two attendants leapt on to their motor scooters and rasped along the towpath to the next bridge three kilometres away like a pair of jogging wasps, past fields where stooks of oats stood drying. The sky was clouding over, and soon stooks, sailors and scooter riders would all need their macintoshes.

A group of gypsy children, along with a barking dog, spilled from a bankside shack; they waved and shouted a greeting, for traffic is rare on this canal. Just before Abbeville we passed the charmingly-named waterside café Au Chant des Oiseaux. The Somme is alive with birds and their song, ducks of many different quacks and colours, coots that chuckle and rummage in the reeds, and blue herons that flap large and lazy along the bank. Sometimes a kingfisher spears the water like lightning, a flash of electric blue.

Soon afterwards we came to Abbeville, which was the scene of fierce fighting in 1940 when a cavalry regiment commanded by a certain Colonel de Gaulle, with the help of a Scottish brigade, held up the German advance. As so often happens in war, the towns that are defended suffer most; Abbeville was heavily bombed by Stukas, and the centre burned. The roof of the church of St Vulfran fell in, but is now restored.

There is a rather grubby commercial basin, but we did not stop. Like so many docks, it is in the less salubrious part of town, and we had no reason to want to pause after only a few kilometres of our journey.

Kilometres do not come easily to a navigator because the nautical mile is a logical natural unit, being by definition a convenient proportion of the earth's surface – one-sixtieth of a degree of latitude. The metre was based on an arbitrary fraction of the earth's circumference, but to the base 10, which does not divide well and makes spherical trigonometry awkward. As if this were not enough, the Paris Academy of Sciences in 1791 did their sums wrong and the metre needed adjusting, and though the standard metre is now fixed in platinum it has rather less logic behind it than the foot, the pennyweight or the troy ounce. (It is worth noting in passing that the learned parish priest of Lyon, some time before, had proposed a metric system based on the nautical mile, but the French Revolutionary Council were not well disposed towards the Church and his excellent advice was ignored.)

Nautical miles have a surprisingly good decimal system underneath them, each mile being divided into ten cables' lengths, and each cable into a hundred fathoms. The fathom is based, like the foot, on primitive body measurements, for its origin is the span of a man's arms, obviously the most practical way of measuring a rope. The measurement is also used by French seamen, being known as *une brasse*.

However, we were not at sea. All data concerning French canals is in the metric system, and while in France we will stick consistently to it.

The river is not navigable through Abbeville, and one has to traverse the town on a liquid by-pass. There is a lock at the start of this, and to enter it one must negotiate both a narrow bridge and a very sharp right-angled turn out of the river into the by-pass canal. With the river current flowing strongly past it is not too easy to poke a long narrow barge into the tiny gap under the bridge. While the fore part of the boat gets itself nicely tucked in and free of the current, one's *derrière* is being pushed indelicately sideways. One feels a bit like an arthritic old gentleman trying to get off a moving train while carrying a ladder.

Just as we entered the lock the gathering clouds broke in a thunderstorm, and Laurel, who was looking after the front end of the boat where there is no shelter, was drenched. Bill stayed dry in the wheelhouse. Skipper's perks.

In the lock, despite the rain, we enjoyed watching the water rising (bearing our boat with it), and seeing the tree-boles come into view, followed by grass and the tow-path. We later found that going down a lock, when the reverse happened, was subtly depressing: the view disappeared, and with it the flickering waterlight that bathes the canal; there was a darkening, and a smell of mud and decay. But we would be halfway through France before we started to descend; now we were at the beginning of a long climb up to the Plateau de Langres, hundreds of kilometres away.

After leaving the lock we were unable to find an adequate mooring, so could not patronise the large supermarket close by. We had a tantalising glimpse of Abbeville station, a classic piece of railway architecture completed in 1912 that Laurel came back later to draw, and at the edge of the canal a monument to a young man who, in the eighteenth century, had been tortured for failing to pay proper respect to a passing dignitary. No wonder the revolution came.

Bill had failed to anticipate the fact that leaving the by-pass would entail a complementary problem to the one already encountered on entering it; we were, after all, both novices as far as canals were concerned. We were rejoining the river at a higher level and it was flowing even faster than at the start of the by-pass.

We had to turn sharply into the strong stream while all our steering functions at the rear end were still held tight in the narrow canal. We helplessly watched our bows being carried the wrong way towards the nearby weir. It was not without difficulty, and a certain amount of damp ill temper that we regained control once again, both of the boat and of

ourselves, and continued on our course for the Mediterranean, sweeping up the assortment of leaves and small
branches that we had acquired during the battle. We felt
fortunate that we had not scooped up a passer-by as well.

Later on, when we had become friendly with some
bateliers, or barge skippers, we asked them if they had a
technique for such junctions. No, they shrugged, one just
gets into trouble, and out of it again. Canals, they added,
were planned by engineers who didn't know anything
about driving barges. Evidently.

We now followed the course of the river, upstream into
beautiful countryside.

France is a much bigger country than Britain, with much
the same population. It has an area of 551,700 square
kilometres, compared with Britain's 228,300, so the
people are spread more thinly. The land seems often
devoid of people, and the villages and hamlets widely
scattered.

Such is the Somme valley. The river flows through
wooded banks, and there is a twinkling chain of small lakes
behind the trees. These were formed by peat diggings, as
were the Norfolk Broads, and it is in some ways a similar
landscape, though more bosky and not so flat. The Broads
themselves are not deep, but these lakes are very shallow,
mostly unconnected to the river, and therefore unnavigable. Both the lakes and the river are a paradise for the
angler, and if anything can be called the French national
sport it is this. Fishing is pursued with an intensity that is an
obsession.

Once upon a time the waters of the Somme were host to
salmon and sturgeon, but no more. Now there are pike
(which the French, unlike the British, regard as a culinary
delicacy) and carp, bream and perch. There are also a lot of
eels. In this area they do not smoke them, a dish we love,
but make an eel *pâté* at Péronne which we looked forward to
trying.

We had expected some ill feeling from the fishermen, because in a big craft one cannot help inconveniencing the bankside anglers, but in the event we found nothing but smiles and waves as we passed. In this respect the Somme differs from the Norfolk Broads, where the anglers do not smile, but there a boat passes every minute, and here on the Somme a passing craft is rare; we were going slowly and carefully, as the river is shallow and its frequent bends can be sharp and impossible to see round because of overhanging trees. We were ever fearful of a barge coming the other way, so our speed remained about six kilometres per hour (about three knots); at that speed one does not make waves to drench the ankles of fishermen, and one has time to enjoy the scenery.

There are other practical factors which limit a heavy boat's speed in a canal, or canalised river. No matter how powerful the engine, and however fast one runs it, one comes up against a definite practical maximum, like an aircraft reaching the speed of sound. Any further increase in power makes the boat not only far more difficult to manoeuvre, but also creates a powerful suction effect on the banks as one passes, thus doing a lot of damage. This is much the same as pushing a cork into a bottle: as it enters the neck there is insufficient space for the escaping air to pass round the cork, and there is a lot of resistance. In boats, one can do damage and waste fuel, but it is also possible to create an impression of brutal power, as the waves overflow the banks, and some people enjoy this, not caring if they annoy others at considerable expense to themselves.

At one point the country to our left opened out into fields and the distant blue of woods and hills at the valley's edge and, in this beautiful setting, amid lofty and ancient trees, stands the Château de Bagatelle, an 'elegant folly of the eighteenth century', as the guide book says.

The canalside villages are pretty and entice one to stop,

but mooring can be difficult, even impossible. The canal is in a bad state of repair and much of the banks are crumbling. With inflexible logic the French government allocates maintenance monies to the canals in proportion to the amount of traffic they carry. The Canal de la Somme carries almost nothing, so it gets almost nothing. Pleasure navigation is free, so virtually the only revenue this canal earns is from fishing licences. This does give the anglers an edge in any argument.

There were *chasseurs* too. We saw one procession consisting of a man in full hunting dress – his breeches and jacket, with 'poachers' pockets', both in a woodsy sort of colour, his hat with a feather in it, and on his back two guns and three fishing rods. Next came his wife with a huge backpack, clearly containing several litre bottles of water and much sustaining fare. The French picnic is not the poor little lunch-box of England: two floppy sandwiches, a wrinkled drumstick and an apple will simply not do if you are as hungry as a French hunter. Then came the dog, of course, lolloping along, unburdened and happy; he seemed to be having the best of it.

There is a lot of game, mostly waterfowl, and at that time the French hunting classes, which means virtually the whole male population, were up in arms (literally) because the EEC had just demanded a curtailment of the French waterfowl shooting season in the interests of uniformity. There must always be a possibility that France will secede over this issue.

In the late afternoon we found a pleasant little grassy quay just above the lock at a village called Long. It was too silted up for us to get alongside comfortably, and we had to walk a long and precarious plank to reach the shore, but it was such a lovely spot that it was an essential stop. Later we would be glad of moorings much less convenient and far uglier, as our journey progressed.

Twosie, after much study, walked the plank gingerly

ashore, and returned in a hurry, seen off by the lock-keeper's tabby.

We went for an evening walk round the village, very pleasant and dominated by a grand *château* of the eighteenth century, now the country seat of a paint manufacturer from Lille. Almost as interesting from an architectural point of view is the Victorian Hôtel de Ville, an astonishing town hall for so tiny a place, with its exuberant belfry on top of a joyous steeply-roofed tower, and green lawns sloping sharply down to the water.

There were some little shops, a visiting cinema, and overlooking the square a small memorial:

> A. F. Hayward
> du Royal Scots Greys
> tombé le 1.9.44 pour la libération de Long

We went into the café-bar for a drink as there was no restaurant in the village. It was very busy. Round the billiard table at the rear there was a good crowd watching the game.

We thought for a moment before joining the spectators, as the last time we had done this in a Yarmouth pub a ball potted by a Glaswegian with more aggro than skill had landed in Bill's beer mug with shattering results, at least to the glass of beer. The Glaswegian had been incomprehensibly and charmingly apologetic. This looked to be a game of entirely another order, and we took the risk. If any apologies were to accrue, we would understand the French rather better than Glaswegian.

Two men were playing: one very *soigné*, well manicured and elegant, with a pink shirt and maroon cords, probably a media person weekending; the other plump and outdoor-featured, presumably a local man. Both were extremely skilled at the game. French billiards, apart from being a schoolboy joke, differs from the English sort by being played on a table without pockets, and the only method of scoring is by what we call cannons – making the cue ball strike each of the other two balls. Bill fancies himself at billiards, but had to confess that he would not have been able to match the exquisite control and high standard of play that we found in the village of Long. No danger of a ball landing in your wine glass there.

We asked the café proprietor about the memorial we had seen in the square. I know not, he said a little curtly, I am not from these parts. A short discussion took place among those who were. The old men had forgotten, and the young ones didn't know. Most of the clientele were young; Mr Hayward's gallantry is commemorated in stone, but his memory has not been passed on to the present generation. It is perhaps like this everywhere: the last war has become something to see in films or strip cartoons, and not something that happened to real people in your own village.

The following morning Bill decided to do some maritime chores, so Laurel went sketching and, as one of her vantage points was near the lock-keeper's cottage, she made friends with the family, for people are always curious about anyone drawing their locality. They were interested in the barge, and that afternoon we were sitting at their kitchen table, elbows on the oilcloth, and being introduced to Marcel and Madeleine, and grandmother O'Lamp (that was what we heard; it dawned on us after a while that her name was Olympe), while the tabby cat made strong efforts to take over the chair Laurel was sitting on. Marcel's father had been lock-keeper before him. Now it was Madeleine who was the *éclusière*, while Marcel supervised the whole stretch from Abbeville to Amiens.

Their cottage was prettily sited with its kitchen garden and chickens on one side of the lock, and a flower garden on the other – a little idyll that is unlikely to pass on to their children if the lesser canals continue to decay at the present rate.

We bought six eggs, new-laid by their white hens. 'What race are they?' Laurel asked. 'Sous Sex,' she was told, which she interpreted as Light Sussex. We tried to buy a cabbage, but they said the cabbages were *pas terrible* (not marvellous), the season had been poor and they would not take any payment for it. Madeleine took off the outer leaves and threw them to the hens, leaving a compact green heart which may not have been *terrible* but did the two of us nicely for supper with some steak.

The next day we enjoyed the almost forgotten delight of new-laid eggs for breakfast. We had found that it was most unusual to be able to stop anywhere for lunch; although the willows and alders beckoned, and the grassy banks looked so alluring, the water was too shallow to approach them. So lunch tended to be something easily eaten *en route*, with fresh bread bought before leaving if we were lucky, while breakfast grew less continental because you never knew about lunch. Well yes, you did know; it was absolutely certain that if Laurel went below and started anything serious for lunch she would be called up to deal with a lock, and that would be that for another half hour or so.

We collected Twosie from her hide under a large purple cabbage, wet with dew, in the kitchen garden; said *au revoir* to Madeleine, Olympe and the tabby cat (Marcel was off in his little orange van, checking his section), and continued upstream.

We looked back and up at the *château* at Long, high above the sloping lawns, its cream stone and rosy brick perched among the green, as light as apple blossom above the river bank. The river was flowing quite noticeably, about one or two kilometres per hour, but navigation was not difficult. We touched the bottom once, but that is to be expected in canals.

At Picquigny there is an unusual double lock, and larger boats have to lock through in two stages. There is a grassy bankside mooring and we stopped for a day or two.

We had left our little car at St-Valery, and as the mooring was next to the railway station it seemed convenient to go back and fetch it.

We got up very early because French rural railways seem to run only when normal people are in bed, and waited on the platform. The elegant station buildings are boarded up and defaced with graffiti; the only sign of former glory is the stationmaster's house with its busy chicken-run and quackle of ducks. Only one other person waited with us, an elderly lady who began to tell Laurel the story of her life. Laurel is used to this; even in countries where she does not speak the language, perfect strangers pour out to her what are clearly the most intimate details of their lives.

The lady was on her way to Abbeville to put flowers on her mother's grave, something she did every week, grumbling about the cost. She could not go every day, as money was short now she was in the *maison de retraite*, or old folk's home. The little outing gave some structure to her life, and something to talk about when she returned.

The train arrived, a twin-coach diesel unit, and we were able to retrace our passage as we passed along the canal banks to the fabulous railway station at Abbeville. Laurel stayed to sketch it while Bill caught the bus to St-Valery. He was its only passenger, as the tourist season was almost over, and the driver made a short detour to drop him just where he wanted near the post office, where there was a big batch of mail waiting for us.

Mail is one of the problems of the wandering life. As nations become more sophisticated and mechanised so the postal services decline. The most efficient we have encountered was a few years ago in Turkey, where, to ease the unemployment problem in a land with no social-security provisions, the government made almost everyone a postman, and letters flashed about the country as they used to in England during the last century.

The *poste-restante* system is not the perfect solution most people imagine it is, owing to the fact that all countries have different filing conventions. As well as checking the 'C's, it is also necessary to look under 'E' (for Esquire), a source of confusion to foreigners and annoyance to Bill, for being a master mariner he has earned the title of Mister, and considers himself rather better than the mere hanger-on of a knight. We also have to look under the name of the boat.

Having recovered the little car, we re-united at Abbeville and did some shopping.

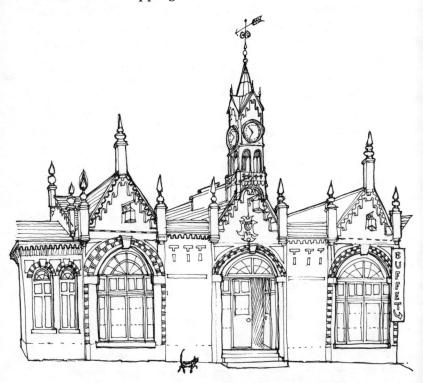

Back at Picquigny we found a rare bird: a *maître gazier*. Yet another shortcoming of the EEC is its failure to achieve a Standard European Gas Bottle. We carry Dutch ones, and they are quite different from the British, and both are quite different from the French. French suppliers will not exchange any bottle other than their own; however, a *maître gazier* is qualified to fiddle about with these things and will for a surprisingly small fee fill any known gas bottle, given a little time. Even with the right bottles the chore of renewing the gas supply is an arduous one, for the bottles are heavy and the agents are never close by. God bless the little car!

Meanwhile our little cat Twosie enjoyed herself in the long grass between our mooring and the railway station, which was overrun with mice. We could watch her leaping about like a feline kangaroo, and hear the unmistakable and rather unnerving noises of her successes. These were not all that frequent, because some of the mice, to her utter astonishment, jumped into the water and swam off – something entirely new to her. She returned on board disconcerted but replete to contemplate the flotilla of young canoeists who surrounded us every day.

Just upstream of our berth was a weir, and part of this had been turned into manmade rapids for canoeists. These friendly and well-mannered children painted the river with their brightly coloured kayaks and yellow life-jackets as they splashed up and down the water slope, cheerfully capsizing and righting themselves under the watchful eye of an adult coach.

We took an afternoon off to see the prehistoric site at Samara, from which it is thought that the Somme takes its name. This has been *aménagé*, which is not always a good thing. The ancient village site is fenced off for the scholars to play in, and the local authority has constructed nearby a 'typical' village of the period through which one can walk. They got a bit carried away and installed in the thatched huts present-day craftspeople who are not all that appropriate to the prehistoric way of life.

As it was the very end of the season, some of the supposed Bronze Age artisans had already gone back to Paris. In the dyeing hut, skeins of coloured wools swung lonely in the autumn wind, the pottery kiln was deserted and a few skins pegged to the tanner's wall drummed hollowly, awaiting the return of activity the following spring.

The enthusiasm of the governing body extended itself to a museum of exuberant modern design containing Madame Tussaud-style tableaux of prehistory, rather too full of messily slaughtered stuffed game. The last of the Parisian Bronze Age artisans looked at his watch as he cast Bronze Age spearheads. There was a small cafeteria. Finally they had run out of prehistoric ideas and added an arboretum, which we enjoyed, and a nature trail. It was quite an instructive afternoon out, and in summer one could easily spend the whole day there. Late in the year as it was, there were still a few families having a picnic among the trees.

Our daughter Shelley, with Nigel the aviator, decided to fly themselves over for a weekend break and, as the nearest airfield was at Amiens, we re-embarked all our bits and pieces, and set off once more.

One approaches Amiens past derelict and unattractive factories, and the quay on arrival is not very welcoming, being about three metres above the water. It has mooring rings at forty-metre intervals (roughly the length of a standard barge), and a large sewer outlet perfuming the only ladder. A four-lane main road passes nearby and the noise was intense. This quay is the Port d'Aval (literally the downstream port), which is intended for *commerce*, though there were no commercial barges in sight.

We passed through an ancient lock under a frightening arched bridge, which mercifully turned out to be higher than it looked, and after by-passing some of the town we arrived at the Port d'Amont (the upstream port), which is being ameliorated for the *plaisance*.

The French have interesting priorities about boats. The

amélioration was not then complete. They had installed neat little cupboards on the quay containing water and electricity connections, but had not yet provided any mooring rings or bollards. Of course no one is going to let that stand in their way, so everyone there was moored to the little cupboards, some of which had not been able to withstand the strain. But it was a nicer quay than the Port d'Aval. Wide, and lined with huge plane trees, the four-lane highway was within a speed limit, a somewhat academic point as most French drivers seem to regard the limit as a sort of minimum, rather than as a maximum. However, there were traffic lights about 300 metres apart, so only the most idiotic drivers, and there were enough to make road-crossing hazardous, could actually get up to 100 k.p.h.

The big trees overhung the cobblestones, and there was a little garden at one end where local inhabitants walked their obviously incontinent dogs, as they do in every well-kept space in France. Even the French are beginning to worry that their beautiful towns are in danger of disappearing under a carpet of canine *crottes* – what we may call the turd world. Apartment buildings are mainly to blame; dogs do not take to the idea of kitty-litter.

Every evening an elderly gentleman drew up in his smart new Volvo and parked under the trees opposite our berth. He would alight wearing his neat suit and dark grey Homburg hat, and while his lady sat in the car he would go round to the boot and take out a bottle and two glasses; he would then rejoin the lady inside, still wearing his Homburg. We would love to be able to report scenes of unbridled lust, but there were none. There was not even much animation. Mostly they sat opposite our windows munching and sipping wordlessly, staring at us as we stared back in retaliation. After about an hour he would pack the glasses back in the boot and depart, to return the following evening and repeat the ritual exactly. Never did the lady leave the car. She was much younger than he was –

what was the set-up? One could write a dozen different stories, and probably none of them would represent the facts at all.

Other frequent users of the quay in the evening were a large number of alcoholics and rough-sleepers, for there were some good park benches in the gardens. A Belgian woman in the next boat had found the launderette full of these *clochards* on a rainy day, and had had to retrieve her still undried washing in great haste when a fight broke out.

Further along the quay there is a big market under the trees each Saturday, with vegetable and flower stalls. Close upstream are the *hortillonages*, the market gardens that have been there since the Middle Ages. Approachable only by boat, the 'paths' between the rows of vegetables are little canals, like a horticultural Venice. As with Venetian gondolas, the boats that serve as gardeners' wheelbarrows are also black, but in the form of a punt with a raised square end, so they can approach the onion patch or weed the carrots. The produce still comes to market in these boats, poled across and down river to the Place Parmentier – what a suitable name for a vegetable market!

Thence, if one crosses the Pont du Cange one comes to St Leu, the old working-class part of Amiens that has been slowly rotting. It is an interesting place with several waterways running through it, and the council has initiated a programme of *embourgeoisement*. Small working-men's cottages are being done up like Chelsea mewses alongside the water, and it has to be said that the general effect is very pleasing, especially the Quai Belu, where the Capitainerie was about to open a restaurant with tables on the quay, an original interpretation of a harbour master's function that would not be likely in Britain.

The river is busy here, but with *plaisance*, not *commerce*. There is a very attractive rowing and canoe club, and an enormous restaurant boat departs and returns twice daily in summer, with much backing and filling.

Once the whole scene erupted with siren-whooping, blue-light-flashing police cars and *sapeurs-pompiers* in red vans, while men in ungainly frog flippers loped along the towpath opposite, trying frantically to find a spot to jump in. We watched a very efficient rescue operation as a body floating in the water was recovered. Doctor and ambulance crew took over for resuscitation, but we could see by the look of despair on the young doctor's face that it was not successful. Slowly and quietly they packed their gear and the cars and ambulances departed one by one, silent now, bumping and backing awkwardly along the towpath. One of the *clochards*, we were told. Drunk or despairing, who knows?

We had not been too happy about Twosie and the quay, mainly because of the large dogs that were walked there. She did not seem to be going far, but we took to calling her in for the night at sunset. We had been invited aboard *Wanderer* for a glass and a chat, so Laurel looked for her as usual before going. 'She just came past you, down the ladder,' said Bill, 'she's under the chest.'

A premonition made Laurel's heart stop. She bent to look under the chest. Twosie lay there, mortally hurt. Our friends on *Wanderer* helped us find a vet, but it availed us nothing; our little cat slipped away from us and died before we got there. Presumably she had been hit by a car.

A very black week followed. As usual when mourning, either for people or cats, disbelief was followed by bitter realisation and a dreadful sense of loss. Acceptance and healing would come, we knew, but slowly.

3

Guests and gamecocks

'E plus de pensamen, qu'es la sagesso de
se leissa pourta sus l'aigo folo, A la gracie
de Dieu, coumo lou ciune En rejougnent
la testo souto l'alo.'
. . . .
'No more heavy thoughts, the wise
let themselves go on the wild waters,
in God's grace, like the swan tucking
head under wing.'

Work and routine is probably the best way to get over bereavements, and we threw ourselves into making the spare cabin fit for the tolerant standards of family, if not quite what royalty would expect. There were no cupboards or hanging rails, but there was a bed. By Friday we had fitted a carpet, a bedside lamp and some heating, and there was even a door for privacy, which was more than we had in our own quarters; we made do with a bookcase baffle that prevented our cabin being totally open to the saloon. The shower in the guest bathroom could be persuaded to work, and the emptying mechanism responded if you kicked it hard enough.

Near the end of the week we tried emerging into the world again, and found the Costume Museum in an elegant private house of the seventeenth century, the oldest still standing in Amiens after the war. Many of the large rooms were given over to a beautifully displayed and arranged collection, changed every six months. We had the place to ourselves, and therefore the individual attention of Madame Carton, who explained the details and history of the costumes that she and her daughter have collected over the last forty years.

The weekend gave a lift to our spirits as Shelley and Nigel arrived. We found a restaurant in St Leu, and had a very good and fairly noisy dinner at La Couronne (Madame didn't seem to share our good humour – we must have been drowning her muzak). On the Saturday we pushed on upriver a short distance, past little riverside cabins for the anglers, and came to the biggest heap of potatoes we had ever seen, by the side of a factory. These were for making *fécule*, or potato flour, which we can strongly recommend for thickening soup and gravy.

We had the usual problem finding a mooring at Corbie. There was a short discussion about the wisdom of running our mooring ropes across the towpath (a dangerous practice, but so often necessary and common on these ill-maintained canals; strictly speaking unauthorised persons, whether on foot or in vehicles, are not allowed on the towpaths, but this is not rigorously enforced nowadays). The problem resolved itself with the news that a wedding party was in progress at a house downstream and, faced with the likelihood of our mooring lines decanting semi-inebriated wedding guests into the canal, we moved on a bit, to find the lock already closed for the night.

The following day, Sunday, we cleared the lock and spent a good deal of the forenoon in the Bar de la Marine, while Bill fetched the car from Amiens. On his return we ordered some *crudités*, which were intended as a little something before going back to *Hosanna* for lunch. Huge and delicious platefuls of raw vegetables with mayonnaise arrived, so colourful and appealing to the eye that it was very late indeed before we got back on board to our Amiens duck *pâté en croute*, roast guineafowl and fruit tart. After that everyone except Laurel fell asleep. No supper required. It all made a change after the snatched lunches we had had *en route*.

We drove our visitors to the tiny airfield at Glisy the following dawn. It was quite deserted except for a distant shape gathering mushrooms into a basket, and a few sheep beyond the perimeter. We watched the little plane take off as light streaked the sky. Strange to think that they would be back in England in an hour – a journey that had taken us five weeks.

We decided that it was time to make some southing, as the old sailors called it. The first leaves were beginning to fall, the swallows were getting restless, and we didn't relish being caught in the north of France for the winter when we had the option of Mediterranean weather.

Quite often the canals freeze solid in mid-winter, especially on the Plateau de Langres, which is the watershed between the northern coasts and the Mediterranean. It is pleasant taking one's time when travelling, but now and then prudence demands a concentrated advance.

One of the principal domestic features of the colder parts of France is the long, narrow pile of neatly arranged logs in every garden, witness to winters that are not warm. Our aim was to head for a region where there were no more woodpiles – like Ulysses sent to walk inland carrying an oar until someone asked him the use of it. In the meantime we set about getting in some coal for our cast-iron stove, the weather having already turned a bit nippy at night. The coal arrived on board in the form of spherical briquettes, which bounced happily around our steel decks like spilt marbles and nearly drove Bill mad. He keeps the upper deck tidy, while Laurel looks after the interior – at least that is the theory. It is, says Laurel, very unfair, because he gets his cleaning done by Mother Nature every time it rains.

It was difficult re-embarking the little car. If we were going to do some real travelling, the chore of having to go back and fetch it at frequent intervals would become intolerable; it was not a car to drive long distances in. The problem was that we had never before embarked it using our own gear. The bits and pieces for doing so were with us, unassembled; they had been heaved on board the day before our departure from Yarmouth, where the little car had been loaded on to our deck with the boatyard crane.

We have to confess that, though the system was basically sound, it had some teething troubles. ('It's a good sort of brake, but it hasn't worked yet,' as Christopher Robin said.) First we had to raise one of our masts, which wasn't too bad. Then a boom was rigged as a derrick and swung out over the quay. The little car had had an eye welded to its roof, and it was easily hooked on; the anchor winch hauled in the rope and we watched our wheels slowly rise to the required height.

Congratulations were premature. As the load left the quay and depended on the derrick it pulled the barge over sideways, so that we heeled towards the quay. To swing the car inwards so that we could put it on to the deck meant turning the whole caboodle on a canted axis, and thus raising the load further. It soon became apparent that the gear as rigged did not give us quite enough strength to do that. Worse, if it had, and we had swung the car across, could we have stopped it before it swung right over, canting the axis the other way to finish up on the other side of the barge, where it would have been suspended over the water?

This is the sort of thing Bill, as a seaman, should have foreseen. Normally he gets cross when he makes a mistake but on this occasion he decided to remain calm. We had got the car halfway in, and there it stuck: we couldn't release the rope hauling it in, and we couldn't haul it in further. It was time to sit down and think.

At this moment, who should appear through the lock but the Belgian family we had met at Amiens, in their extraordinary boat which they had built themselves in a rather narrow corridor at the side of their garage. One never criticises another man's boat, in the same way that one does not cast aspersions on his wife or children, but finds something to be complimentary about; let us just say that a boat shaped like a corridor did not appeal to us. 'Very original,' we told him. Whatever the shape of his boat, he was a man of a practical turn-of-mind and helpful disposition.

We didn't have to explain the problem. Four backs are better than two, especially well applied, and in no time our little car was settled on to the top of the *roef* by main force, where Bill had calculated that it would give us about ten centimetres' clearance under the lowest bridge on our route, a margin that could be expected to decrease as we used up the fuel in our tanks. Careful calculations about such technicalities as waterplane area had convinced him that we would rise about one centimetre for every ton weight that we shed.

We checked the stores and made them up from a simple supermarket. We took a last walk past the odd-looking abbey church of St Pierre, built between the sixteenth and eighteenth centuries, and then truncated of two-thirds of its length because the fabric threatened to collapse. It has an impressive façade and then peters out; it now looks like the bust of a dachshund cut short just behind the shoulders.

We had a cheerful dinner ashore at La Table d'Agathe, and were then ready to resume the great adventure.

We have written before about canals designed by people who do not understand boats. There is another example a little way upstream at Froissy. On the downstream side of the lock, just before the gates, the overflow from the higher water level is released back into the river at right angles to a boat's line of advance. When it rains the sideways force of water is so strong that it is virtually impossible to enter the lock in an orderly fashion, and this provides much innocent amusement for the otherwise bored agricultural workers in the village. Such chutes are called *diversoirs*, and occur occasionally, which is far too often.

This is the part of the country where the battle of the Somme was fought in 1916, though the main fighting took place away from the river. It was characterised by monumental incompetence on the part of the British generals. The British High Command allowed the Germans plenty of time to dig in on the higher ground; our soldiers 'going over the top' to attack were not allowed to run or charge, but were ordered to advance in line at a walking pace towards the machine guns, the better to be mown down. The Commander-in-Chief, Earl Haig, ignored sound advice from the field and refused to attack at night (Henry the Fifth won at Agincourt, and he never did that – why should we start now?); the newly invented tanks were used for the first time, but the recommendations of their inventor as to their best use were ignored, so their huge potential was wasted; and finally, when at enormous cost some ground was taken, the Command dithered and lost it again.

Silent witness to the resulting carnage is still there in the vast war cemeteries, where the white crosses walk just as the generals wanted their soldiers to walk, in neat rows among the poppies.

If the main action was to the north-east of the river, evidence of the battle is still to be seen along the banks. The river was the main supply route for the British, and this was the busiest period in the canalised river's history. Rails were laid alongside the canal for little engines to tow the barges that supplied the army; and these rails can still be seen, rusting and drooping as the ballast has been long ago washed out.

It would be easy to underestimate the effect this battle had on Britain: we found British people who still made an annual pilgrimage to honour their fathers, uncles and other relations, often several in the same family. They would see our British flag and come over to talk to us.

Wars have been fought on French soil for centuries. In

1346 Edward the Third of England won a little battle against his cousin at a place called Crécy, which gave him the throne of France. Crécy is north of Abbeville, and as we passed the area Laurel made *potage Crécy* with the superb carrots of Picardie. It was a much better soup than we had supposed, and a most attractive colour. Picardie is renowned for soup.

It is also noted for some of the ancient country sports. One evening, out for a stroll and still a little way from home (by which we mean *Hosanna*, of course), we stopped at an unpretentious café-bar in a tiny village. The building was one of those hundred-year-old houses, with tall narrow windows surrounded by a checkerboard pattern of bricks, that abound in Picardie and the North. Inside the slightly forbidding entrance there was a cosy little bar with a high oak counter from behind which *la patronne* welcomed us cautiously. The only client was an old man in the corner straining a *bock* through his walrus-moustache.

After the ritual greetings we obtained our drinks, and during the conversation we asked if there was a restaurant nearby.

'No, there is not one restaurant hereabouts,' replied *la patronne*, 'but if you would like to, you could have here a very good repast.'

'What do you have?'

'There is a good pot of vegetable soup, perhaps some veal chops garnished, and the farm to the side has a fresh cheese of the most excellent.'

'That sounds very good,' we said.

La patronne paused only to charge the old man in the corner with looking after us and, donning her coat, she pedalled away westwards on her bicycle.

The walls of the café were decorated with the sporting trophies typical of most small French cafés. We had not given them more than a casual glance, expecting to find the pennants, photographs, and regalia of village football. But as Bill took the old man another beer (at this point it was we who were looking after him) we saw that the photographs appeared to be of chickens.

'It empassions you, the combat of cocks?' asked the old one, showing his first sign of real interest, apart from a flicker as the beer arrived.

'We don't know much about it. It's forbidden in England, though it still occurs.'

The old man mumbled something that might have imputed a lack of manliness to the English, but was more likely to have been the after effects of a *bock* of French beer.

'Do you have cockfights here?' we asked.

The old man thawed. 'Not here, but one will find them. City people,' he added darkly, 'they don't understand the country and its patrimony.'

'Shame,' we said. The three of us paused to consider the shortsightedness of city-dwellers.

'Come,' he said, rising with a little difficulty and picking up a stout stick. He led the way to the back door, and into a garden about thirty metres by ten, with all sides except the house lined with neat white painted cages, all at about chest height. They were of four different sizes and there were about a hundred in all.

The aged one explained that cocks came in four weights, rather as boxers do, and that the little bantams were the most savage, a characteristic that is common in humans, too.

He stopped in front of a big cage, the occupant of which was obviously dear to him, for he looked steadily at it while talking quietly to us. This huge gamecock was, he said, very ferocious. It was a true champion of all the champions.

The bird eyed him through the wire netting with undisguised hatred, its neck feathers ruffled. It was a big golden cock, its comb clipped to a minimum to afford an enemy no purchase, and as it turned in its agitation we caught a glimpse of iridescent black and green tail feathers.

The old man cooed gently to the bird as a mother might talk to an infant. The bird, we thought, remained unimpressed. Then the aged one, displaying remarkable courage and agility, opened the cage door and thrust in his right hand at floor level as if to grab the bird's legs. There was a feathery flurry in the cage, the old man withdrew his hand sharply and a few loose feathers followed it on the evening air. He uttered the *mot de Cambronne*, and sucked the blood on the back of his hand.

'Oh, this is a bird of the most ferocious,' he said loudly, while said ferocious bird shifted from one foot to the other and put its head on one side to bestow on him a look of ineffable wickedness.

If we expected to move on at that point we were mistaken. The French peasant is not that easily beaten. The old one talked gently and lovingly to the cock, stroking the top of the cage with one hand. The cock watched the deceiving hand, hungry to get at it, showing thereby why cocks fight to amuse humans rather than vice versa, for suddenly the unwatched hand opened the cage and the old man dragged the protesting, squawking, flapping, feathery mass into the open, and without a pause whirled it round and round overarm like one of the more aggressive West Indian fast bowlers limbering up before eating the England tailenders.

After two or three loops he brought the unfortunate bird to a halt at arm's length, holding it firmly by the ankles, while giving it a couple of sharp slaps to its cheeks. The bird seemed to have temporarily lost its belligerence, not to say its sense of well-being; it now looked more like a vicar who had just done a full round with a heavyweight champion, and was trying to decide if he still believed in God.

The old man cooed to the bird. 'Oh, how he is beautiful,' he chanted, 'truly champion, champion of all the champions, my beauty, my true beauty.'

We began to feel nervous about the bird's imminent and

likely recovery from its ordeal. That the thought had also been on the old one's mind was obvious, for as the bird began to get him in focus again it was unceremoniously shot back into its cage and the door slammed and securely fastened.

The old man sucked his bleeding hand once more as we walked back to the café. Words seemed a little superfluous. We bought him another beer, and he chuckled happily as he exchanged healths with us, and then *la patronne* returned.

'Don't believe a word the old one says,' she warned as she hurried into the kitchen. Her voice floated out: 'He is a teller of little stories, that one.'

The old man winked at us. 'Women are all the same,' he sighed.

The dinner was simple but very good and surprisingly cheap. The fresh farm cheese, eaten with a little caster sugar, was delicious and creamy. But the cabaret had taken place before the meal.

We do not give the whereabouts of this little place, though we were made welcome, for they do not seek visitors. But people wishing to see a cockfight, which is still quite legal in France (how could it be otherwise in the Land of Logic when the Midi has its bullfights?) can go to the little village of Tourmignies north-east of Douai. There at the Gallodrome there is a well-built cockpit, measuring three metres by four, like a boxing-ring with rounded corners and a gate at each end. It is raised about a metre from the ground, and close by is a pair of scales for the weighing-in. Round the cockpit are tiers of seats for the spectators, with capacity for about 250 people. The cocks here are kept in large hutches, clean and well-maintained, while the females run loose in a grassy yard. Anything less like the comfortable plump domestic hen, your Light 'Sous Sex', or Rhode Island Red, can scarcely be imagined. These rangy ladies reminded us of Olympic runners – lean, hard, and with long muscular legs, though instead of an athletic tan they sported pewter-coloured scales.

We did not stay for a fight, as the temptation to experience something unusual gave way to a feeling that the performance would not be very edifying. We chickened out, you could say.

During the next few days, as we travelled, we saw that the poplars and ash trees were beginning to take on a faint tinge of gold and shake a few leaves down; the woods at the water's edge had an untidy look, the rosebay looked tired and shabby, a toss of columbine was tangled in the brambles, and there was a general air of dishevelment.

The haws were thick and scarlet, while the wild briar's hips of golden-orange were more delicate and less lavish. It was a day of showers, ripples and rings on the water, and a strong rustle in the trees. Some blackberries had forgotten the date and were in flower. A coot stood still on the bank and hoped we would miss him.

After Sailly Lorette, where the view opened out on the left to rolling corn stubble, beanfields and maize, three little boys shouted, *'D'ou venez-vous?'* We called back, *'Angleterre!'* They waved to us and ran on.

We could see further now, to the little hamlets and a distant stand of dark trees, all the same height and with trunks as neat as a row of railings.

Just above Cappy lock and before the little lift bridge, the canal takes a gentle left turn, and on the right bank there is a comfortable mooring under some magnificent horse-chestnut trees. We took advantage of it. This tiny village had four bars, all inexplicably closed, and a very expensive gastronomic restaurant, which was not closed but was out of reach of our pockets. A white hen, scratching in the road outside the church, suggested that eggs could be bought at the farm.

After Cappy the canal became even more wild and even more narrow. We grew a little apprehensive, as great trees overhung the navigation, often with their branches sweeping the water, and woolly hats were at risk from acquisitive twigs. Bends became very sharp, and two barges would have had difficulty passing each other. Our upperworks were continually brushed by the branches, and once more we ended up with leaves and twigs all over the deck.

It was ladies' day, we found; they dashed out to open bridges in their aprons, they operated the locks and a large woman at Frise in a lilac cardigan bent high above us at the lock's edge to take our line, revealing vast but very discreet undergarments. Only the very young girls wear trousers in Picardie. At the edge of the lock one of the black punts from Amiens had become a flower bed filled with geraniums.

We discussed the whole subject of canals as we bumbled along at a peaceful four kilometres per hour. Their *raisons d'être*, their history. How odd they seem now, but how vital they once were.

The problems of transport that our rude forefathers had

to overcome must seem almost unreal to the footloose flyabout of these days. We have nowadays a choice of how to go, how to carry, when to go and when to tarry. Life is easy, if a bit more complex.

It is hard to realise that until well into the reign of Queen Victoria transport throughout most of Europe depended on the strong backs of the horse, the ox, the mule, the donkey and the walking feet of the packman. The only exception was water transport.

The proportion of trade carried by coastal shipping was astounding. It was far quicker to carry your coals from London to Newcastle (assuming you were daft enough to want to) by brig than to pound your way over the rough cartroads of the period.

The difficulties of long-distance trade were reflected in the tendency for land-locked communities to be largely self-supporting and introspective. The Romans changed all that. Heretofore, when tribes had a fight the victors would seize what they could carry, and after some carnal celebrations would return home. The Romans had the novel idea of administering the lands they conquered, and to do this they had to improve transport.

Such is the intellectual insularity of terrestrian historians that we hear much of the Roman roads, built for the fast travelling of trained soldiers on foot and horseback, but little of the fact that these roads were on the whole inadequate for carrying goods.

Few people understand the importance of military logistics. Though you could force-march troops a long way over clear straight roads, it was difficult to supply both the troops and the increasingly elaborate administrative tail that accompanied them, with the lumbering carts of the time. They even took with them materials for the mosaic floors considered necessary to underline the standing of Roman dignitaries (rather as the size of carpet designates one's rank in the Civil Service). The craftsmen needed to make them went too. Heavy carts pulled by heavy horses over roughly-paved roads made for a slow voyage.

France has always been an important route between Northern Europe and the Mediterranean. Further east the continental climate produces very hard winters that close the mountain passes, whereas France is to some extent helped by the same oceanic influences that keep Britain warm(ish).

When the Romans wanted to expand northwards they used the river Rhône, or Rhodanus, as they called it, to carry their goods. We can get some idea of how desperately they needed such a route by recalling some of the difficulties of navigation on this mighty fast-flowing river before its recent canalisation. Frequent floods, and especially flash floods rising within hours, and a normal current downstream of rather more than walking speed, made it a hazardous and difficult route, but it must have been a lot easier than their marvellous roads. Downstream on a big river must have been the express delivery of the day.

Later they went on to use the Rhine in the same way. Big rivers, turbulent and troublesome, but they were there for the use of the bold and determined. The problem was how to cope at the points where river traffic became impossible or unmanageable. So it was back to backs – horse, mule or human – but it was a slow process because the loads that each unit could carry were so small.

The Romans finally hit on the idea of building some canals, but they could build these only across level ground. Unfortunately, no one had yet discovered the lock. The world struggled on in this lack-lock state for some centuries, while emperors and empires came and went.

It is not known for certain who invented the lock. A pity, because, after the wheel, it must rank as one of the most useful and simple mechanisms ever devised; and as with the wheel, one wonders why no one thought of it before. There is no firm evidence that it was the Dutch, who are sometimes given the credit for this. With the flat nature of their country, locks would not have been so vital to them anyway, though it is possible that a lock of a primitive sort existed at Damme, near Bruges, in the twelfth century. But as the Dutch word *damme* is the origin of the English word dam, it was probably a very primitive one indeed. Whoever thought of the original idea, however, a lot of the kudos for its practical development goes to two Italians, the Signori degli Organi and di Bologna, who actually constructed a working lock at Viarenno, near Milan, in 1438. Leonardo da Vinci (who else?) built six more round Milan in the 1490s. The first lock in Britain was created at Exeter in 1564.

The first French locks are thought to be those of the Canal du Briare, connecting the Loire and Seine rivers, in about 1640. From then on one loses count.

The French canal system differs from the English system in that its shares were not bought up by the railway companies, as happened in England. The English canals were therefore starved of new investment by the railway companies who owned them and who were their chief competitors; they were never enlarged as the Industrial Revolution advanced, and were effectively wiped out by their owners. In France, by contrast, the whole system was modernised and enlarged in about 1878, when both the railways and the canals were nationalised under the Gambetta government. Oddly enough, the English mistake is being repeated today in France, where the SNCF, the national railways system, is being heavily subsidised while the canals are being starved of funds. The French railways, with their powerful trade union, have a lot of political clout, while the barge crews, mostly small family concerns, have almost none at all. As trade dies, so the barges are sold.

Pleasure navigation on inland waters has also not developed in the same way in France as it has in Britain. The Norfolk Broads, for example, had a multitude of cruising boats for hire a hundred years ago, and these companies expanded on to the British canals soon after the Second World War. In some places they have now saturated the waterways. It was one of the Norfolk companies, Blue Line, which first introduced cruisers for hire on to the French canals, mainly because they had little room for expansion in Britain. Since then pleasure navigation has developed further, with other British and several French companies moving in, and the waterways are now increasingly being used for this purpose.

The Canal de la Somme, which we were travelling, has no commercial traffic these days, and it was while we were debating this sad state of affairs that we came out into the Canal du Nord, swooping out of the overhanging trees into the daylight like leaving the Tunnel of Love at a fairground. For the Canal du Nord is different from the Canal de la Somme. It is part of the *grand gabarit*, a network of canals that were enlarged to take bigger barges. There is an even grander *gabarit* envisaged which would match the German, Dutch and Belgian systems, but plans are not far advanced, for the reasons we mentioned above.

So, suddenly, we had left our little farm track and were in the middle of the M1. Here we met the big boys for the first time: double barges, totalling seventy-six metres in length, with full-sized cars on board, and domesticated wheel-houses with polished brass, potted plants, lace curtains at the cabin windows, and washing flying. Despite our worries about meeting a barge round the corner on the Somme we had not as yet even seen one, and now barges were going both ways at what seemed to us a terrifying speed. Some of them were roped together in tandem, driven from the back of the rearmost barge, where visibility must be difficult, to say the least. Suddenly life was faster, more dangerous; we needed a sharper lookout for threats to which we were as yet unaccustomed. But it wasn't far to the town of Péronne, and there we moored for the night at the commercial quay, our little barge dwarfed by the big *péniches*.

On the opposite side of the canal, by the big bridge is a café-bar much frequented by the *bateliers* (the French barge crews). Apart from the usual refreshments it sells rope and other nautical hardware, and repairs two-stroke engines. Among the bar tables is a work bench full of useful-looking metal bits and pieces, oily rags and evil-looking chains.

The establishment is kept by a large and intelligent dog, which occasionally allows its owners a say in events. If they assert themselves and shut it out, it will fling open the door with a reverberating crash, come in and take charge once more. It approaches each customer as they enter, to approve or otherwise. If the latter, it sits in front of the embarrassed client growling softly and continuously. Its name is Pollux; Bill asked them to say it twice as he misheard the first time and remarked, 'How appropriate.'

We were approved by the dog, who got his people to serve us a couple of drinks and lend us a torque wrench. We were escorted to the door when we left by the tail-wagging *patron*.

It was a long walk into Péronne town, but a very pretty one, past the wooded *étangs* and the allotments on their banks. It was a nicer town than we remembered from a previous visit, when we had had an appalling meal at a local hotel known to us ever after as the Place with the Black Staircase. This sombre décor, combined with the French stairlighting system which is programmed to go out just as you are halfway between light switches, led to Laurel being benighted on the wide spiral staircase in total darkness and having to yell for help.

This time we found a *traiteur* of outstanding excellence. The French *traiteur* is the origin and the epitome of the 'takeaway', and his cooking is often far better than that of the restaurant chef. Sometimes it is more expensive, too.

We ate very well that night, *chez nous*: eel *pâté*, a speciality of the region.

During the next day we had less wind, and even the aspens were almost still. The swallows dipped and swooped, coming south with us.

Locks on the Canal du Nord are double size, and mechanised, so one passes through quickly, and in no time we reached the junction to re-enter the Canal de la Somme. Care was necessary not to miss it, for coming off a farm track into a motorway is easy compared with finding the farm track in order to leave one. The entrance was unsignposted, virtually concealed behind a large fallen tree, and looked like the sort of backwater that is soon only a couple of inches deep and choked with rotting barges. But it wound gently on, free of impediments save for the pendant boughs. We were at slow speed once more, dodging the anglers and the trailing branches, watching the tufts of reeds bow their heads and dip under the water in the ripple of the wake like shockheaded boys forced by Matron to wash their hair.

Picardie is a generous land. For 7000 years it has been cultivated by succeeding tribes, and has been for centuries the granary of France. It is the biggest producer in France of sugar beet, peas and potatoes, and still the second biggest producer of corn. Thus our nightly mooring was quite often a silo quay where once these products were loaded into barges. Now the factories remain at the water's edge, but everything comes and goes in huge lorries which encumber the roads and foul the air with petrol fumes, and kill people who drive in little cars, and have the temerity to display stickers saying: 'If you have it, a *camion* brought it to you.'

The wild wooded banks gave way to the potato fields of Ham (pronounced 'Arm), where on the outskirts we passed the biggest sugar factory in France, about to open its doors for the start of the sugar-beet harvest, that period when to travel by road is to run the gauntlet of the beet lorries,

always overloaded, bouncing in the potholes that orna-
ment the country roads of France and spilling their cargo
like cannon balls on to the road. '*Danger, betteraves,*' the
notices say, and lucky is the motorist who reaches his
night's lodging without signs of bombardment by the
dreaded *betterave*.

We entered the small lock at Ham, the gates so out of
shape that water poured through and over its upper doors,
roaring like a waterfall. As the barge rose we found the lock
surroundings neat and well kept, and a friendly young
éclusier, who gave us good advice on where to moor, helped
us make fast to the best mooring we had found so far. He
had little trade, we discovered, only two or three barges a
week, and he spent his time reading in his hut, listening to
pop music or gardening in his plot.

Ham could not be called inspiring. One supposes that a
town which once contained one of the most notorious
prisons in France is bound to have an atmosphere of
depression. The copperworks had closed, the restaurant
had lost its star, aerial postcards showed nothing but the
beet factory. The less said about its railway station the
better. We were moored opposite the ruins of the old castle,
once the prison where, amongst others, Napoleon III
had been held until he escaped and fled to England for
sanctuary.

We stopped for a few days to do some work in preparation for winter, the quay being just the right height to act
as workshop. Bill's gooseneck had been bent when hoisting
in the little car, and he took it to the blacksmith to be
welded. We found a source of logs for our stove, good hard
hornbeam, about a ton of it, which we stowed in a neat row
along the deck.

Bill took the diesel stove to pieces. Cleaning it, as any
woman knows, was a process of transferring the black oily
smears first to himself and thence to the cleanest of the
galley towels, which, as every man knows, is a more

effective dirt-absorber than the Black Watch tartan rag placed there for his exclusive use.

We knew that there was no point in going on too fast as we had heard that the lock at Berry-au-Bac would be closed before we could get there. The junction at Berry-the-Ferry is one of the busiest, where traffic from Belgium joins that from the North, which has in turn recently joined that from Paris. An interruption here was serious; there would be many barges held up, and it didn't seem wise to get too close too soon.

In France there is sometimes a major *foire*, for which whole towns are closed to traffic. There was one in Ham, and we spent half a day wandering round in spite of cool showery weather. We looked for a shop that had tickled us in the past: Ham Radio. It had become a travel bureau with a boring name. The local Rotary Club were offering *moules-frites* cheap for a pleasant lunch.

We bought at a stall a piece of lace curtain with water birds on it – just enough to cut into three and give a little privacy to the windows on one side of our saloon. Looking for some garden canes to thread them on, we heard news of a kitten in search of a *foyer*; the lady from whom we bought the canes offered to telephone.

Yes, there was a kitten, female, white and tabby. We fell for it, and walked to the house of a thin widow, who explained that *un chat sauvage* had given birth in her attic, an event that must happen quite often in Northern France, where the attic windows always stand open, if indeed there is any glass in them at all. We have tried to find out why this should be so even in the depths of winter, and have for once encountered total incomprehension. (But all the world holds open the window of the attic!)

Our widow did not discover the illicit kittens before her dog, who ate a couple. The one she showed us now was the survivor, how could she keep it when her little Tou-tou was so jealous? 'Is the cat a good mother?' we asked, catching a glimpse of the wiry black cat in the road. It seemed so.

The kitten was about a month old, its nose torn by its encounter with the dog, wild, angry, afraid and spitting. It stank enough to make even those with the most insensitive noses reel back several yards.

We couldn't possibly.

We did. Subject to a clean bill of health from the vet next door. We put the little treasure into a shoe box, and took it to the vet.

Another couple were waiting their turn, in the tiled airy room. Stacked everywhere were jeroboams full of horse mixture, and vaccines for cows, to be administered a gallon at a time by a hypodermic syringe the size of a fire extinguisher. Everything was on an enormous scale, as might have been expected in an agricultural area. A Large Animal Practice, we thought.

Oh dear.

We looked at the other couple. They were elderly and rotund, and their kindly crab-apple faces, both fairly whiskery, were bent in ineffable love over a black rabbit with very long front teeth. The tenderness of the scene reminded us of a Flanders School Nativity painting. The rabbit had been a beloved pet for a very long time, they told us. Every month they placed it, like Ernest, in a leather handbag with handles to it, and brought it to have an inch or so lopped off its front teeth. It was not painful, they said, he leaped *tout seul* into the bag. Otherwise, he would arrive not to eat.

We felt reassured.

When they came out beaming, the rabbit wrapped in a baby shawl and its nose wriggling with joy above its newly shortened teeth, it was our turn. The vet was very young and very good. He did a few fast and unprintable things to the kitten, who replied in kind. He said it was female, slightly less than a month old, healthy and needed worming.

Of course.

He gave us some powdered MilkoDogue, with instructions and an eye dropper, and we told the widow that her troubles were over and ours were just beginning.

Bill came down with a cold at this point, and other uncomfortable symptoms conducive to inactivity. We had a discussion as to whether it was not better in a foreign land to go to a vet rather than a doctor, as the former were used to diagnosing their patients without language, but, as Laurel pointed out, one ran the risk of being wormed every three months.

A day or two later, when Bill had recovered without needing further remedies than those we carried with us, we moved on and spent a few quiet hours' travel, uneventful save for watching the waterlily leaves. As we passed, they rolled up and dived under water, folding like a closing fan. It made compelling viewing.

At Jussy we came to our first descending lock, and a short stretch where the canal was above the surrounding countryside; we could look across open countryside until the woods closed in again.

The locks at Fargniers were the sort with chickens and ducks and some other exotic-looking birds. A goat was tethered close to the towpath, and two sheep on the opposite side. When we arrived at lock number 3 at Fargniers, a notice was quickly flapped out of the lock office window. '*Vente de lapins*,' it said, but we were too busy weaning an opinionated kitten to be bothered with rabbits. Meals become simple when there is a lot going on.

So we came to the canal junction at St-Simon, where we joined the Canal de St-Quentin and the Freycinet-standard commercial canal system. We left the peaceful Somme behind us, clothed in the green and gold of early autumn.

4

Mainly Aisne

*'L'aigo esbrihaudo e ris; li pesqueirolo, li
barbasan, fasènt l'aleto, Rason l'oundo
fugènto au gai soulèu que viro.'*
· ·
'The sparkling water laughs, the
coots and swallows glide a-wing,
skimming the wave that flees the
wheeling sun.'

We were by now north-east of Paris. We wished to by-pass
this city, not out of any disrespect or disregard for its many
attractions, but because we were not at this time seeking the
populated part of France. We have often been to Paris; like
so many tourists we have climbed to the top of the Iron
Lady, smiled enigmatically at the Mona Lisa (hearing the
inevitable know-all behind us hiss: *'Ce n'est pas le vrai!'*) and
been totally terrified by taxis and traffic. We love it (one of
us more than the other) but not at the same time that
everyone else wants to love it.

So we preferred to swap the mighty roar of city life for the
mighty roar of combine-harvesters, the growl of tractors in
the potato fields, the bellowing of cattle and the demonic
shriek of jet fighters practising to save us all from a fate
worse than deafness caused by the noise of low-flying
aircraft.

Going the long way northabout Paris involved dodging
among a network of waterways, a few kilometres on this
canal, a few more on that river, then a junction, a spell on
another canal, and so on.

Our plan from the end of the Canal de la Somme at

St-Simon (shortly after Ham) was to navigate the Canal de St-Quentin for all of eighteen kilometres to Tergnier, there to fork south-westwards for about twelve kilometres to Abbécourt, where we would once again turn through ninety degrees to the south-east into the Canal de l'Oise à l'Aisne.

All that in one day! We spent the night at Guny, where a wakeful cockerel at least suggested that there was a *basse cour* with hens where eggs might be obtained next morning. There was indeed a henrun, and we called at the farm. The farmwife had, she regretted, only six eggs to sell (the hens were off lay) and no bags to put them in. Bill brought them home tenderly in his pockets.

That night the clocks went back an hour. One is not always aware of this event, which is in France a closely guarded secret. We had consulted calendars and diaries to no effect, so we went on to ask the Gendarmerie, several French citizens and the Mairie – all of whom also consulted their calenders and diaries, to no effect.

Our uninformed guess was confirmed by the radio the following morning. There have been times on our travels when we either guessed wrong or forgot all about it, lived happily out of step all Sunday, and found out only on Monday, usually by turning up at the shops an hour before they were open.

The following day we enjoyed the rest of the long trundle of forty-seven kilometres to Bourg-et-Comin, where we turned left again into the Canal Latéral à la Marne, which would be our route for a further twenty kilometres to Berry-au-Bac, where another junction awaited us.

And that was enough for the moment, because we knew that the lock at Berry-au-Bac was closed for repair. It was a scheduled *chômage*, or stoppage, one that could be planned for. Some months previously a Dutch heavy-lorry driver, perhaps subconsciously aware of his country's nautical heritage, had decided to return to Holland via the canals

instead of the highway, and had chosen the bridge over the lock at Berry-au-Bac to make the change.

Unfortunately, he was no better at handling his improvised ship than he had been at driving it as a lorry, and he failed to bring his suddenly converted vessel to a halt before hitting the lock gates. Not surprisingly, this had caused an *arrêt de navigation*. Whereas a *chômage* is a planned stoppage, the *arrêt* is due to an emergency. Makeshift lock gates had replaced the damaged ones and were operated by hand until resources were gathered for the main repair, which was now taking place. Berry-au-Bac is a point where three important canals meet, and the effect on barge traffic was something like that of a derailment at Clapham junction. There was a certain amount of disorder, and some grumbling of a resigned and rueful nature, because heavy trucks are killing the canal traffic quite fast enough without resorting to physical violence against lock gates.

Thinking that there would be a queue of commercial

barges waiting for the reopening at Berry-au-Bac, and aware that not only would we have to give them precedence out of courtesy for their livelihood, but that all the convenient moorings would be taken up, we ambled lazily in that direction, planning to arrive at Berry-au-Bac two days after the re-opening, by which time the log-jam of *péniches* would have dispersed.

So we pottered along, pausing to listen to those birds that were left after the French *chasseurs* had passed, and watching a little diver bird with a long beak. And of course we tried to eat out, as the galley was still a makeshift affair. Anizy-le-Château looked good. There were little symbols in all our guides (except the Michelin Red) telling us that there were good restaurants in the town. The absence of a mention in the Michelin Red Guide should have told us something, for Mr Bibendum doesn't miss much. But one always hopes to be the first to discover some cheap and perfect little place that he has not yet found.

Anizy-le-Château seemed a nice little town. Many of the smaller places in France have not only their proper name (i.e. Anizy) but an additional noun as a qualification. A classic example is De Gaulle's home town of Colombey-les-deux-Eglises. To the north of us in the coalfield is Bully-les-Mines, nearby is Sin-le-Bon, which needs no comment, and after Long we had come to Longpré-les-Corps-Saints, a name we savoured, thinking of the Sainted Bodies brought to the Long Field after the Crusades.

One of our self-imposed duties when travelling in France is to provide towns with imaginative suffixes when the forces of history have neglected one of the chief 'dwaddlums' conferred on Frenchmen by the revolution: to live in a town with a suffix. As two public-spirited and generous Britons we do our best to remedy the deficiencies of the French Constitution where appropriate.

Ham-le-Sandwich is obvious, and we claim no feat of imagination for it. Picquigny-les-Deux-Pissoirs was one of our better ones, being nicely alliterative. We imagine it to be the birthplace of our archetypal French impressionist painter, Saturnin Pissoir, who was always on the point of joining Van Gogh in Arles, but never quite got there.

What, you may well ask, is a dwaddlum? This phrase was on the air many times a day at that time; the French being currently obsessed with the coming *bicentenaire* of the revolution, where they used to think dwaddlums originated until Mrs Thatcher went to Paris and 'spat in the soup'.

Spelt non-phonetically it is written *'droits de l'homme'*, the Rights of Man. The French revolutionaries certainly set out some 'rights', which notably excluded women from all benefit (note that the women of Turkey got the vote before French women), but then their Declaration was based on that of the seceding colonials in America, who excluded black slaves and 'merciless Indian Savages'. No Bill of Rights, or Charter, or List of Freedoms – call it what you will – is perfect.

Anizy-le-Château is notable at first sight (to canal folk) for its fine new supermarket on the banks of the canal. There are not many shops on the canalside, and it is often something of a trek to obtain provisions. A shop close to the bank is to be profited by, as a chance to buy the heavy things. It is salutary to remember the days of one's childhood, when potatoes were lugged home on one's bicycle only if the grocery cart had failed to call for a few days, and the Corona Man would bring the Dandelion-and-Burdock to the door and collect the empties. The butcher and grocer delivered; milk, bread and fish came to your door. Nowadays everyone does all that heavy work themselves.

There were two quays next to the supermarket; one of these had sloping sides and would have made a very uncomfortable bolster for *Hosanna* to lean against, while the other had lost its mooring bollards. However, there was a tree, and though the rules of navigation forbid mooring to trees we notice that all the barges do it when convenient bollards are lacking, so we follow suit. It does not seem to do the trees any harm; some of them have clearly been used this way for many years.

So we were able to push our trolley-loads of beer and potatoes and five-litre 'cubes' of wine right down to the boat. Obviously the bargees did the same, for there were several supermarket trolleys abandoned, either awash in the water, or upended on the bank. We stopped there two rainy days and observed why the trolleys were not returned, for barge people are normally very orderly.

Because of the lack of bollards, the barges moored with only one rope and left their engine going ahead to hold themselves in position against the bank while Madame went shopping. This isn't the safest way of coping in a narrow canal, and the pressure was evident to get under way again as soon as the goods (and Madame) were back on board.

We suggested to the *direction* at the supermarket that they would lose fewer trolleys if they could persuade the authorities to replace the bollards, a thought that had not occurred to them.

After spending a small fortune in order to eat over the next few days, we went in search of the high life of Anizy, which seemed to be divided into two by the canal. The western, lower part had a railway station, an apparent restaurant which turned out to be a café, and a signposted, advertised restaurant with gourmet pretensions; this had been closed and turned into a night-club.

The steep climb to the older part of town brought us to an attractive area with several good shops and a little hotel with a menu which completely failed to generate any water in our mouths at all. We decided we would do better *chez nous*, especially as the larder had been restocked. We had a small wander round the town, and in a side street were rewarded by the discovery of an Aladdin's cave.

This was a *brocante* – the largest and most comprehensively combined junk and antique shop we had seen for a long time. It occupied an enormous stone barn and was stacked from floor to ceiling with articles reflecting the rural surroundings and the lives of the people of Anizy.

There was an artistry in the way in which ancient tools and farm machinery were hung or disposed about the floor and sorted by kind: rows of hayforks led an attack across the wall, encouraged by serried lines of sickles and billhooks, halting only when confronted by a determined assortment of frame saws in different sizes. Bicycles dating from long before the war hung from the lofty beams, dipping in a draught, their handlebars locked, discussing old times. In one dark corner a gaggle of sewing machines sulked under their wooden covers, and in another stood stark machines whose use is now almost forgotten: chaff-cutters, binders, butter-churns and pumps, and a peculiar mangle with curved rollers that we cannot imagine a convincing use for.

On dusty trestles, under oily paintings of a dark, religious nature, were bakery moulds, mincing-machines, copper pans, biscuit-tins, tea-caddies, candlesticks, broken objects made out of a thousand matchsticks, and lamps made from wildly inappropriate objects. A line of crucifixes crossed one wall, contemplated by a chipped plaster saint of indeterminate sex, and at the darkest end of the barn several Gargantuan wardrobes, big-bellied, hunched their shoulders and lowered at the smaller furniture. Wicker baskets spilled faded curtains; assorted boxes offered sepia postcards, sheet music, old hats, old handbags, old shoes. Laurel was so fascinated by it all that she came out, blinking, without buying anything, but Bill found an old copy of the Michelin Guide for 1926 and a folding art-deco wall lamp. Laurel could be heard muttering: 'Where bicycles hang by the wall, and choppers hang in rows on nails . . .'

On our walk back to the barge we came across a recently

dead mouse, and Laurel wondered whether to take it back for the Treasure, like a good mother cat.

Bill discouraged her.

The Treasure, who we now called Tassie, had cost us some sleep while she was being weaned. We had shampooed her on returning from the vet with something kind to kittens, and dried her with the hair-dryer, which she enjoyed; so she smelt better, and turned out a lot whiter than we had thought, with silver grey tigerstripes. Part of the smudge on her nose turned out to be built-in, and the fleas were only slightly diminished, as what is kind to kittens seems also to be kind to fleas. A few of them fled, probably with soap in their eyes.

We mixed up the MilkoDogue as instructed. She scorned the eye-dropper given us by the vet at Ham, and went straight for the spoon, splashing and sneezing, but keen, every three hours. It was not so much that Laurel fed her on demand during the night, (3.30 a.m. seemed the favourite time), but that she learned almost at once to climb up the bedspread. Even when, at the end of the week, she was sleeping through the night in her own box, she was prone to early waking and instant boredom. This meant that she would wail a bit; then, when nothing happened, she would come to the three steps down to the bedroom (barges are not quite like houses) and wail a bit more. After about three days she learnt to fall down the steps if nobody answered, and then it was 'yoo hoo! party time!' with ourselves as unwilling guests, as she roared about the bed and bit things. We began, perforce, to get up a bit earlier, especially as the end of the day was darker now.

The swallows were still with us as we continued our journey, though September was nearly over. The grey-green smoke of distant willows in the mists of morning looked like a Corot painting, but a painting in constant movement.

There seemed never to be a still day as we travelled,

always a wind blowing in the tree tops, always a ripple on the water. The woods were wild and untidy, the fall of the leaves revealing the white bones of dead trees and broken branches under the spread of bramble and convolvulus. The poplars and willows were patched and tasselled with mistletoe, and every morning the hedgerows were covered with Spider lace, hung there in the autumn sunshine for the dewdrops to dry. The tips of the poplars were turning gold, like a dipped paintbrush, and the chill of the evenings led us to light the wood stove and put the winter duvet on the bed.

The duvet was not so close to the floor as the bedspread, and for a few days we had a respite from Tassie's morning games; but she soon mastered the Climb and thought up some more: Moling and Burrowing, for example, or the Attack on the Second Foot. When all of us were exhausted she would fall asleep on Bill's neck. If he dared to move incautiously, she bit his ear.

We got under way once more towards Braye, where this canal reaches its highest altitude. It is this point which presented the canal engineers with a major problem. Every barge crossing the watershed will cause two locks full of water to be lost from the summit pound. On a busy canal this can be a considerable amount per day, and the problem is that water in quantity is not easily come by near hilltops. When planning canal routes of this kind, the engineers have to locate the highest water source of sufficient quantity for their purpose, and make the summit pound a little below that.

In order to even out seasons of little rain, or periods of intense barge activity (Monsieur Murphy would obviously see that these coincide) the engineers usually provide a good-sized reservoir. On this canal, where the highest source of water was not as near the watershed as they would have liked, they had to lead the canal through a long tunnel. The reservoir just before the tunnel has been made into a leisure centre, with all the aquatic toys one could wish for in very attractive surroundings.

It was this tunnel of Braye-en-Laonnais that we were approaching in some apprehension, as it was our first such experience. Fierce notices added to our disquiet:

'Do not stop without warning between Lock 9 and tunnel.'

'One way only.'

'Respect the traffic lights.'

'Respect the middle of the tunnel (dangerous banks).'

'Possible wait one hour.'

The dark tunnel arch appeared before us, and we entered from the 'Oise' end. Nowadays, when canal traffic is a fraction of its former business, one is allowed to pass through the tunnel at Braye under one's own power; not so long ago one would have had to join on to the end of a tow of eighteen or twenty barges.

The tunnel is 2365 metres long, and there is a traffic cop at the southern end. He controls the traffic lights which tell you if there is anyone coming the other way, for it is impossible for two boats to pass in the tunnel.

The passage takes about an hour; we were lucky, there was no one in sight, the light was green, and we poked our nose between the walkways at either side of the arched vault. These walkways seemed ideally placed for keeping the high sides of the wheelhouse clear of the lower part of the tunnel wall. Under these circumstances we felt we could increase speed a little, for we were almost exactly the same width as the channel, and it did not seem as if we could come to any harm. Though the tunnel was supposed to be lit, most of the lights were out and, as the daylight dropped astern, there was a dank smell and a noise of echoing drips from the vault above us.

We drove on in stygian gloom, our small spotlight seeming to have no impact whatsoever.

Oh, hubris! With only a few feet of warning the walkway on the right came to an end. Sixty tonnes of barge does not lose its momentum very quickly, nor is it easy to steer when slowing down. Within half a minute we veered over to the right side and winced as our starboard navigation light, which is on the corner of the wheelhouse roof, lost an argument with the stone arched tunnel.

The noise was appalling. Not the little rasp and tinkle you might think would accompany the disintegration of a plastic object containing a light bulb. No! A screeching, grating, long-drawn-out, reverberating, crunching sound, with an obbligato of tinkling glass, a pattering rain of small pieces falling on to the deck, and a few small splashes, which might have been Bill's tears of shame for such an awful bloomer.

We were now illegal, for the rules clearly stated that navigation lights were obligatory in the tunnel. The water cop at the far end, if he looked along the tunnel from his high observation platform, would have seen the extinguishment of our bright green light. How could we look him in the eye as we went past, emerging at the 'Aisne' end of the tunnel, feeling him sneer like a Petty Officer instructor of the old school: 'We hit the side, didn't we? What are we? We are a silly sailor are we not?'

Or would he shrug and say, '*Quel con!*' which means, 'What a——' – and is very much what we thought of ourselves too.

Our subterranean adventure had taken fifty minutes, Tassie had hidden down in the hold, the canal cop had waved cheerily as we passed, and we started to descend the downhill chain of locks. So trim and tidy the canal was now, with neat grass edges and hedges cut back, the machinery smart with paint, and a tarmac towpath. No longer the wild tangle of alder and dogwood, exuberantly encroaching until the canal was merely a barge-width between two opposing boughs.

Everything here knew its place, and so must we; the next four locks are all linked and automated. Once in the chain you are not allowed to stop for any reason other than an emergency. If you stop you activate men in little yellow vans who dash about the countryside pushing and pulling, with great heaving sighs of resignation, the levers that release you. Happily we avoided this shame, and at the junction of Bourg-et-Comin we joined the Canal Latéral à l'Aisne, adorned with enormous sugar-beet factories.

We were approaching Berry-au-Bac now, and wished to find a stopping place for the three remaining days of the *chômage*. We chose the little village of Maizy, where there was a charted quay about thirteen kilometres short of the hold-up at Berry. The quay turned out to be another of those sloping stone banks which are not very comfortable to lie alongside when the wash and suck of passing traffic throws you against them. But here there was little traffic, owing to the *chômage*, and the canal opened out into a wide basin.

As we have said before, the massive drag and thrust of a barge in a narrow canal is caused by the wall of water pushed ahead of it. This water cannot get past the barge as it travels, so the level builds up. The barge sucks along with it an area of lowered water level, and it is the contrast between these two levels, which can differ by as much as a foot, passing within seconds, that makes one's moored boat surge. If one is not moored and has to pass a barge going in the opposite direction, this also has to be reckoned with. As the barges' sterns pass, their screws, thirsty for water, suck it in from either side.

Even though barges slow down when they pass in a narrow canal (and one that does not is very unpopular), their sterns attract one another by this suction and their haunches frequently bump and give a *coup de fesse*, or boomps-a-daisy. Where the canal is wider, there is room for the water to get past the barge and the difference in levels is less marked. So the most comfortable moorings are where a canal widens out.

We lurched about a bit at Maizy until we improvised two planks to shore us off from the stone quay wall. It was a wide grassy quay lined with stone houses. Between the houses and their barns were lanes going up to the roadway about fifty metres away. The place seemed deserted, except for the crew of the barge with whom we shared the quay. They kept themselves busy painting their already immaculate boat, high out of the water with no cargo.

That evening we went to the village café-restaurant (Le

Rivage) for a drink, in the faint hope that the restaurant might be open. It was. We had a convivial evening in a bar that was very like an English country pub in an agricultural village, except that this one sold newspapers and fishing tackle as well. Apart from the language, Bill could have placed most of the customers alongside the regulars of the General Wolfe Inn at Laxfield in Suffolk, which had once been his home for a short time.

A man came in and, overhearing our conversation, remarked to the proprietor and the company in general, 'I commence to be an anglophobe.' He looked our way.

Bill said mildly, 'That exists in France? I do not believe it.'

The anglophobe joined in the ensuing roar of laughter and bought our next drink, and from then on we were welcomed into the company.

The budding anglophobe was not so in fact. He was very much *en colère*. Problems with a rented property had caused some sort of family upheaval and made him angry, but as he regained his composure he became very good company.

The conversation was mainly agricultural, reflecting the comings and goings of village life, so that we were not able to contribute all that much. But there was also a good deal of politics: not domestic French politics, from which we deliberately exclude ourselves, but European politics, where the iniquities of those people in Brussels are the common subject of discussion.

Most people seem to feel that the Common Market is good in principal, but not as it affects them directly. The ordinary person in many walks of life is feeling neglected and powerless to influence events.

However, there were plenty of cheery topics, discussions about weddings, the imminent start of the champagne *vendange*, the merits of the local band, and even the state of the *navigation*, for there were some barge people there too.

Growing hungry, we asked if the restaurant was indeed open. Yes, Madame was cooking for the *archéologues*, of whom there was as yet no sign, and would be pleased to cook for us too. Our request to see the menu caused a moment's pause, but a few minutes later it appeared, written in biro on a paper napkin. It read: *jambon cru*, *quiche*, *veau à la crème*, *fromage*, *tarte*. We felt even hungrier. It was now quite dark, and nothing much seemed to be happening in the kitchen, so we listened to a pensioner recount the tale of the time he was 'dead' for three days, collapsed after vomiting a lot of blood. We felt a little less hungry. 'The doctors told me: your liver is fine,' he said, sketching with his hands a healthy sized organ, 'but you must stop smoking. That made me think: I am not ready to die – and I have not smoked since,' he said proudly, walloping into his sixth glass of *rouge*.

A van hooted outside, and a weatherbeaten farmwife entered, in men's corduroys. She was *la laitière*, and came twice a week with the day's milk, eggs, cream and bowls of fresh white cheese. She collected half a dozen mineral-water bottles, and returned with them full of milk. She then brought in a round tin tray on which she balanced a large quantity of eggs, just managing not to roll them. As we had been imprudent enough not to bring our containers to the pub, they lent us a Perrier bottle for the milk, we bought some of the cheese, and Bill walked back to the boat fifty metres away to put them in the cool.

The ingredients for the *quiche* had arrived, as had the archaeologists, and our hostess, Irène, set to work. We were bidden to table to choose our wine and eat our smoked ham with crisp gherkins and good bread, while the smell of baking wafted appetisingly from the kitchen.

The archaeologists were visually indistinguishable from the younger farm labourers; although their Paris accents would no doubt have been obvious to the French, we are not as perceptive as that. There is a major Roman site not far from Maizy but, as in many other instances, money is short for a proper dig, and the enthusiastic professors and students were camping out on the site to do the work on a shoestring.

The quiche, made from that evening's eggs and cream, was placed before us, cheesily steaming and delicious. Now we understood that dinner time was not governed by the clock, no *heure militaire* here. Cooking started the instant the absolutely fresh ingredients arrived.

The main course was a *sauté* of veal with rice and a rich cream sauce, followed by the fresh farm cheese and then a fine *tarte maison*.

The whole five-course meal, with a bottle of wine, coffees, and four liqueurs came to eighty francs each, less than eight pounds. It is this aspect of eating out in France which is so appealing. There is expensive pretentious fare to be had, and there is mediocre food at mediocre prices, but it is possible to find, both in the towns and in the countryside, restaurants providing for the needs of people of moderate means at prices that are astonishing to Britons. The quality is often equally surprising, as this one at Maizy turned out to be.

There are quite a few of these reasonably priced restaurants with good, simple, *table d'hôte* menus in France. Their proprietors make a fair living in a country where the taxes and the cost of food are similar to, or higher than, those in Britain. It is not the same by any means all over the continent, but is a peculiarly French phenomenon. Long may it continue. It makes one wonder how British caterers can look anyone in the eye with a clear conscience.

The next day we heard that the *chômage* at Berry-au-Bac had turned into an *arrêt de navigation*. In other words the repair was being extended by another week, and we were therefore effectively confined to Maizy for several more days. We prepared to enjoy it.

We spent some time painting the boat, as all good bargees do when stopped for any reason. We also spent a lot of time in the warmth and friendliness of Le Rivage, or sitting outside it on sunny days talking to Irène. This late in the season she cooked only when she chose, which was usually lunch for a village wedding, or a weekend party of huntsmen, or, exceptionally in the evening, for us and the archaeologists. At other times she would take off the clean apron and come and talk food, or other interesting subjects. She mourns that the young of today are forgetting the good old dishes, such as the *pot-au-feu*, or the *petit salé* (salt pork) with cabbage, and is delighted when her *pot-au-feu* disappears to the last morsel and the plates are wiped clean with bread. There is, she says, a *snobisme* which regards dishes where the vegetables are mixed in with the meat as lower class.

She herself has never tasted Jerusalem artichokes, as she was brought up just after the war. Many French in wartime had lived on little else, and her mother could no longer bear the sight of them. Many of us war babies feel the same about Spam. With other vegetables, however, her mother was very inventive, adding the few scraps of meat that came their way. Irène rejoices in tasty but economical cooking, and it is what her clients expect; she crosses the road to the farm for the day's vegetables and salad, and the egg lady calls. She cannot be doing with fast food and factory-made dishes whipped through a microwave oven, though we noticed that she has one, and uses it with discernment.

The village is almost completely rural: farm-workers, a few factory-hands at sugar-beet time, no weekenders from Reims or Paris here. There is no hotel, not even a *pension*, and no *épicerie*. There is a baker; the butcher and the *épicier* come in separate vans several times a week, hooting, and the aproned ladies come in twos and threes to the summons. An old lady wishes to give an order for Friday; the *épicier* promises to deliver. 'But,' says the old lady, 'I shall be at a wedding. How shall that arrange itself?'

'No doubt there is a little corner of your courtyard where the things could be placed in your absence?' says the *épicier*, glancing through the wrought-iron gates to her house.

'Indeed yes,' she says. 'I will therefore hold attached the dog on Friday.' There is a solution for everything.

There are many vegetable plots in Maizy – everyone's garden seems to have a *potager*. For a while, some years back, prosperity reigned in the village, and the plots were neglected, but now that there is a good deal of unemployment the plots have returned to favour.

Not only agriculture is in recession. We were introduced to Ginette, a one-time *marinière*, or bargewife, who now lived in the village. She has buried two husbands and two daughters, but her weathered face is calm and her smile charming. When her first husband, a farmer, died, her second took her barging all over Europe, Holland, Belgium and Germany. The children went to boarding schools, which were expensive, she said, and it was heartbreaking to part with them for three months at a time. One of her children died falling into the hold of the barge and breaking her skull. In their heyday the barges carried gravel, coal, beet, iron and grain, but the traffic has declined greatly. Most of the barges we saw on this stretch were loaded with sand for the St-Gobain glass works. Lorries are more adaptable and easier to unload; how many times we hear this, but so seldom the opposite argument, that barges can carry thirty times the amount a lorry can, and do not jam up the highway. Nor do they use up so much of the earth's energy resources. For instance, one horse power can move:

by road	150 kg.
by rail	500 kg.
by inland water	4000 kg.

If one takes into account the full costs of maintaining the route, then the approximate costs of water: rail: road are in the proportion 1: 1.5: 6. The distortion arises because roads and their tributary accesses are maintained not by the users of the vehicles that run on them, but largely out of general funds. But the motoring vote is powerful; no one understands or is bothered with canals, though canal transport seldom kills, which can hardly be said of the roads. And just think of the savings in energy and pollution.

Irène has three big sons, and a little girl. An afterthought, we suggest? 'No,' she says cheerfully, 'I changed husbands,' and Jean-Claude, lanky and bearded, grins. He smokes heavily, as do all his clients; we returned from a session in the bar with our clothes reeking. Preoccupation with diet foods and non-smoking is still a long way from rural France.

Three jolly *dames* stopped Laurel on the way to the baker's, hailing her with some joke at her expense. They were mortified to find that she was English, and that they might perhaps have offended, but in the ensuing chat all was explained. It was her 'third leg', or walking-stick, that caused the merriment, since, not being tired, she was marching along gaily, swinging her stick. The same three ladies walked past the barge occasionally, and continued the chat.

On the Sunday Irène had an anniversary party to cook for, and lunch for the huntsmen. Men in breeches, khaki boots, and strange headgear unloaded guns from the boots of their cars and plunged into the woods, from where, in due course, came a barrage of gunfire, the shouts of beaters and the barking of undisciplined dogs.

'What do they hunt?' we asked. Not a lot, said Irène, they have already killed everything. Except that this year there are too many deer, so some culling is allowed. Irène is fond of animals: she has eight dogs and a dozen cats, all called 'Minou' – Puss. When the inevitable kittens arrive, they are exhibited (at their most appealing stage) in the bar, and all find good homes. It is considered very bad luck to drown them once their eyes are open. 'How can you kill a kitten who can look at you?' says Irène.

On Monday afternoon, a quiet time for Le Rivage, we piled into Jean-Claude's middle-aged Renault, depredations courtesy of any or all of the eight dogs, and were taken on a tour of the countryside. We spent a pleasant afternoon, never having realised that there could be so much of interest in such a small part of the Champagne district. The *vendange* had started and we went in search of a relative who was doing casual work in one of the vineyards, only to find the management rather surly, and not prepared to allow the slaves even two minutes' chat. 'At so many sous per hour you rest in your own time, not mine! *Au travail! Pas de repos! Et vous . . .*' and she indicated that the further we went, and the faster, the better pleased she would be. The *vendange* lasts ten days, and one supposes that every second is vital, especially in unsettled weather. We left feeling a little unsettled ourselves, guilty even, caught committing an offence against commerce in the Champagne fields. We are, however, not able to boycott this house's wine, for we find that champagne producers do not as a rule grow their own grapes, but buy them in from small *vignobles* such as the surly lady's, so we do not know for sure where her grapes go each year. In any case, we don't drink all that much champagne.

The set-back was minor. Other places were most friendly, and one must expect people preoccupied with the climax of months of work to be a little grumpy. We discovered that most of the grapes going into champagne are dark-skinned (Pinot Noir) and not, as we had expected, all white grapes.

We went off to visit the Chemin des Dames, a long bridlepath along the crest of the watershed we had tunnelled under, which had been frequently ridden by Mesdames, the daughters of Louis XV in the good old days. We were also introduced to the area round the village of Craonne, which is strictly outside the Champagne (*appelation contrôlé*) district, so that it cannot by law call its wine champagne. The wine is made in the same fashion, but without the cachet it sells much more cheaply, and many of the people in the trade believe it to be better than that of the lesser 'real' champagne houses, though no one would compare it to Krug or Mercier.

At the close of a perfect day Jean-Claude treated us to a bottle of Castellane outside a small restaurant, Le Postillon in Ville-en-Tardenois, overlooking the Marne valley. There we watched the sun go down and enjoyed the wicked feeling of drinking champagne at teatime; even the toddler had a taste between gulps of some violently raspberry-scented drink. More up-market than Le Rivage, Le Postillon too is an excellent restaurant, with very good cooking at a moderate price.

On Monday the barge traffic started to pick up, a sign that the work at Berry was over. We invited our hosts to accompany us on the short voyage to Berry, which would take a little over two hours. Though Irène had been born in this village and her house backs on to the canal, neither she nor Jean-Claude had ever done the trip. The canal is a closed world known only to those who gain their living on it; it is quite common for someone to live all their lives by the canal bank and never once go afloat on it.

So Jean-Claude and Irène embarked, we were blessed with another fine October day, and although there were storm clouds they seemed only to intensify the golden light on the maize, the sunflowers and the corn stubble. As Jean-Claude steered the boat, Irène sat with the Treasure on her lap. Tassie was unusually biddable: she tried one tentative bite and got her nose boffed. *'La maline!'* exclaimed Irène fondly. They told us about the floods that cause havoc in the river Aisne, which here flows alongside the canal at a lower level. During one flood the river rose higher than the canal, inundating the countryside, so that a barge lost its bearings and ended up stranded in the unnavigable river. It is probably still there, waiting for the next great flood.

We arrived at Berry-au-Bac after tea and scones and entered the repaired lock garnished with neat flower beds, and a *bureau de contrôle* where all traffic has to 'sign in'. Here we bade our friends *au revoir*.

We went to look for some supper, but the restaurants were closed.

Well, it was Monday.

5

Mostly Marne

'Tranquilamen, au fiéu de l'aigo bello, li barco descendien, ribejant d'isclo.'
....

'Tranquil down the calm water the boats descended, coasting along the isles.'

In the event we would have found plenty of mooring space in Berry-au-Bac as it was provided in the days when there were ten times the number of barges that now exist, but we were quite content to have done our waiting at Maizy. Berry straddles an important main road, and its principal products seem to be noise and dust.

We handed in our name, nationality, cargo, port of origin and destination to the *bureau de contrôle* at the first lock, and imagined a *fonctionnaire* in a dusty room in Paris riffling through the papers and saying, 'Tiens! *Hosanna* has reached Berry-au-Bac!', then moving a miniature Union Jack on a map of the waterways and going home to his well-earned *pot-au-feu* in a little apartment near the Porte de Bagnolet.

On the other side of the first lock at Berry the canal opens out into a long narrow basin. Here empty barges waited for cargo contracts, trade being very poor. We took a sharp right turn into the next section of canal, de l'Aisne à la Marne, nearly overshooting it, not having appreciated how close it would be to the first lock. Had we missed it, we would have been in the Canal de l'Ardennes, and on our way up to Belgium.

We moored at a quay belonging to a silo, about 100

metres above the second lock. Normally speaking this would have been a bad mooring, the canal being narrow at this point, but we were close to the operating *bâton* of the lock. Downstream traffic passes under a wire stretched across the canal, from which depends a long pole, or *bâton*, reaching down to about a metre above the water. As the barge passes, the bargee grabs the *bâton* and gives it half a turn, and thus sets in motion the operation of the lock before he gets there, saving a lot of time. In order to make sure of reaching the *bâton* he approaches at a slow speed, and so doesn't disturb the water. The mooring was consequently comfortable.

Although we had been unable to eat ashore at Berry, we gave some custom to the little combined grocery, hardware shop, and bar 'de la Marine' next to the second lock. This is one of the really old-fashioned types of village grocery, whose products are geared to the needs of the canal folk, and we were able to buy the makings of a simple supper. The hardware department, in the same small space as the grocery, sold those items appropriate to maintaining a barge: paint, detergent, rope and so on.

The bar was interesting; the main clientele were barge folk and their dogs, and it was difficult to decide if the humans were more diverse in their appearance than the animals.

Conversation was canal-oriented, with much talk of who had passed by, or had passed on, and in what direction, and what likelihood of cargoes there was. In those parts a barge that had finished a working voyage was then waiting about twenty days before getting another, and this bar is one of the clearing houses for gossip on the canals. The atmosphere of the place was depressing. The customers and the proprietor were polite and good mannered, but they all had the air of waiting for someone to die. Not a soul laughed the whole evening. Trade, as we said, was poor.

As we passed through the ensuing chain of locks, the weather turned stormy, the sun disappeared, the wind rose and the rain tipped down. By Loivre a full gale was blowing across the canal and we had difficulty getting into our berth on the windward side. An unladen barge had even more trouble in getting into the lock.

When loaded, the deck of a barge is not many centimetres above the water and thus presents little windage, but when light, or *vide*, two metres of slab side are exposed, and there is correspondingly little of it in the water to provide a grip. As a result a strong wind will blow it all over the place like a paper boat on a village pond. We turned out in the rain to help this hapless *marinier* and, though we had acted partly in self-defence, we received his grateful thanks.

When a light barge leaves a lock in a cross wind, it does so at full speed, because only when it has gathered way is it under any sort of control at all. This can be really frightening to those moored close by, but the bargemen are professionals and generally know what they are up to.

However, like eminent surgeons, they occasionally make mistakes.

When a laden barge leaves the lock, it has no alternative but to leave very slowly, even though its engine is working flat out. The barge is designed to fit the cross-section of the lock as nearly as possible, and there is only an inch or two all round for the water to get behind it. When it finally emerges it accelerates quite sharply, but not to any great speed. The same problem arises, if to a lesser extent, in the canal itself, and experience has shown that the maximum operating speed for a laden barge in a canal is about six kilometres per hour. Above that the fuel consumption rises dramatically without any real effect on the speed. The extra energy all goes into wave-making, and breaking down the banks.

We needed foul-weather gear for the evening tramp to the restaurant at Loivre (closed), came back to our barge and steamed gratefully by the wood stove, eating a supper of leftovers. Although we travelled on, the weather was terrible for a few days and we endeavoured to stay 'indoors', except for locks, as the first real cold of winter arrived. A window left open had soaked our canal guide, and Laurel had to iron all 289 pages of it to make it usable again.

After many weeks of travel, one settles into a routine. Our day, though busy enough, was not nearly as arduous as that of barges carrying cargo: they would be waiting for the locks to open at dawn and drive hard all day till the locks closed after sunset.

We would get up at about eight-thirty, start the generator, cook breakfast, and heat the water; about half an hour of obtrusive noise would follow, particularly noisy because the generator was not yet sound-proofed. When we switched off, peace returned for only a short while as the main engine was then started, to be warming up while we retrieved our ropes and cast off.

How far we went in a day depended on how we felt, the attractions or otherwise of places ashore, whether we needed supplies, and the possibility of moorings. These we learnt to grab if we saw a good one, even at about two or three-thirty in the afternoon. If we passed a likely spot relying on promises in the guides that there was a better one round the corner, we were usually deceived. Very often the recommended spot would be too shallow for anything other than a rowing boat.

One cannot turn and go back in a canal, for even a small barge is longer than the canal is wide. Turning is only possible at special turning basins, and many of these are silted up. Reversing for any distance is difficult and conducive to bad temper. So we sometimes ended up wet and tired in the gathering dark, with nothing to do but stop and hope for the best; Laurel often had to throw herself into a blackberry bush at dead of night, in search of non-existent bollards.

Mostly, we stopped early, lit the fire, and went exploring while there was still a bit of daylight; the nights were now drawing in.

We did not stop at Reims, as we both know it well. Both of us like to look at good stained glass, so we have made numerous stops at Reims in the past, to see both the mediaeval and the Chagall windows. Now we were on a canal trip, and in search of peaceful France. This is not to be found at the moorings in the centre of Reims, close to a roundabout and an urban motorway.

The canal through the city is interesting, however. There is a big factory making steel cable on the northern outskirts; this provides a lot of work for barges, which are loaded with the huge drums on which the cables are wound. Other steel factories also engender traffic which is oriented more towards the north than the south. Thus there are twice as many barges each year using the canal northbound from Reims as there are southbound.

Unexpectedly, part of the canal through Reims is quite narrow, and as there is a busy rowing club, whose members sit in their boats facing the way they have already been, there are some exciting moments, which are apt to occur just as you want to gaze up the Rue Libergier to the cathedral, hoping for a glimpse of the sculptured Smiling Angel on one of the porticos of the façade; a magnificent view that only lasts a few seconds.

It is virtually impossible for a barge to succeed in attempts to avoid a rowing boat, the latter has inevitably to yield. If they are not looking the way they are going, which is almost always the case, it is worrying for those in the bigger boat, and it requires a stern frame of mind to be able to plough on regardless in a heavy barge. These close encounters, while avoiding boredom, also increase the tension, but you cannot compare it with the North Circular in rush hour, or even with the Avenue General de Gaulle, which crosses the canal close to this point, choked with hooting cars.

The condition of the canal deteriorates after Reims, and we were soon in another tunnel, that of Billy (no relation). This tunnel was in poor shape: the towpath on the right-hand side was dangerous in places, but only half the lights were out and we made a much better job of getting through this one. We came out into the sunshine with relief after an hour underground, and found ourselves in another of the downward chains of locks which carried us on remorselessly, hurrying us through the rolling slopes of Champagne. The town of Bouzy is off the canal to the right; here they make a still champagne of high quality. It is always pleasing to contemplate a glass of the appropriately named Bouzy Rouge.

At Vaudemange the lock-keeper's index finger, wagging like an inverted pendulum, told us off firmly for using old car tyres (as all *plaisanciers* do) as fenders to absorb the rubs and shocks inevitable in locks. Their use is forbidden unless one has an inflated inner tube inside them. This is because an old tyre which has sunk to the bottom of a lock can jam up the works and be difficult to remove. Most barges do not use them because there is not enough space between a standard barge and a lock wall to take a motor tyre, even that of a Mini. But they are much used by the *plaisance*, because the fancy little balloons sold in yacht shops do not stand up to the wear in the canals.

In fact motor tyres are almost the perfect fender – flexible, cheap and long lasting – and in the canals the majority of lock-keepers turn a blind eye to their use, especially if the tyres are firmly attached with two lines and you are diligent about retrieving them if they do escape.

In the earlier stages of our voyage we had asked a few lock-keepers about the use of tyres. It is formally inter-dicted, we were told, but of habitude no person will say anything. The lock-keeper at Vaudemange did say some-thing, but he was not unfriendly about it. We pulled our tyres in for a while, and then tentatively used them again next day without any further query.

If we had been continuing to the south in haste we would have turned left at Condé-sur-Marne, but we had decided on a short diversion to call on a friend who worked at Epernay. We didn't plan a lengthy interruption to his livelihood, but he had told us that there was an excellent mooring in the town, so we turned right.

Epernay was too far to reach the same day, and we had a typical experience of not stopping where we found a good mooring, and being trapped in full flight. Condé is not the most attractive of sites, but it does have a good quay. Feminine nose wrinkled at the prospect, and masculine determination wavered. (It always does when feminine nose wrinkles.)

We went on. A little further there was yet another silo making a lot of noise and dust. Finally we read all the guides and decided that there must be a mooring at Dizy, the last lock before Epernay; all the books said so, and all the maps showed it. On the way we came to Tours-sur-Marne, with a Logis-de-France restaurant alongside the lock. We passed it. We must have been mad.

After Dizy town the canal deteriorated, banks crumbled and there wasn't the semblance of a quay to be seen. Eventually we reached Dizy lock after dark; the lock would be closed of course, but we thought that there must be something there.

There wasn't. In pouring rain we attempted to moor, with the kitten, having suddenly mastered the steps up to the wheelhouse, excitably underfoot. Both banks were completely unapproachable, and we stuck on the bottom some five metres out into the middle of the canal with no chance of making fast to anything – both of us wet, tired and irritable, and Laurel coming down fast with a cold. We restored our spirits with hot food and hot toddy, but spent a worrying night.

The following morning started badly too. The kitten had, with difficulty, been put into the cat basket before we got under way, as we had learnt last night that she was now beginning to get in the way in a crisis, and every lock is a potential crisis. As we entered the lock, dreadful wails were heard and we found that, unbelievably, she had her head stuck through the wire mesh at the front of the basket. One of us had to cope with the lock alone, while the other had to stop the Treasure from strangling herself until help came. When safely out of the lock and under way assistance arrived with a pair of pliers, and released the kitten from imminent garrotting. The accompanying tart exchanges can well be imagined. Who was where when they were needed, and by whom? And who should have known what, and who could go to hell, accompanied by who else, and what else could have been done, and whose fault was it anyway?

Laurel's cold did not improve; not only the weather was

chilly that morning, as we passed under the big bridges into Epernay.

The grassy river banks here would make attractive moorings, though as the river is inclined to flood, some care would be needed in their use. But nobody seems to have thought of this. They are all very busy making champagne.

We met our friend at a small private mooring that was as good as he had promised, with water and mains electricity alongside us.

We also had a close view of the exuberant Art Nouveau mosaics on the sixty-metre tower of the Maison du Champagne Castellane, saying 'Budapest', 'New York', 'Helsinki', and so on. Laurel has had to do with decorating buildings. It is an exercise in the art of the possible. All you have to do, she discovered, is cut down the client's grandiose ideas to what he can actually afford, and then get it passed by the committee. She envisaged the client, de Castellane, asking for pictures in mosaic of every capital where Castellane was drunk, and the artist saying: 'My life, already; you want it done *this* century?' The happy idea of putting up the names of the cities in letters occurs to somebody, and the artist goes off to re-do the designs and the estimate, muttering: 'It should look like a Metro station yet.'

This time we did not attempt the 239 steps to the top. In fact we didn't do much at all; the sun had come out again, literally as well as metaphorically, but Laurel's cold subdued her social life, and as it was Monday not much was open. She cooked an enjoyable British curry (if that isn't a contradiction in terms) for our friend, who was greeted by tiny Treasure with such deep-throated growls that we began to wonder if there wasn't some Rottweiler in her make-up. When she had finished making her opinions clear we were allowed to converse, and at last we were able to find out for sure, from someone who worked for the company, that the final 't' in Moët and Chandon *is* pronounced. Apparently the original Mr Moët was of Dutch origin; it is not a French name.

Curries, like stews, ought to have been made yesterday. It is no use, Laurel says, expecting ingredients who have only just been introduced to settle down together immediately; this only seems to work with fish. An odd expression, 'cold fish', since a fish is far from stand-offish, and mingles quickly with its companions – wine, butter or just lemon juice. Not so the meaty components of curries and stews, who take a day to get to know the vegetables, herbs and spices before they exchange flavours.

After a couple of days we resumed our southing. The river was wide enough to turn in and we retraced our steps to the junction at Condé, where we re-entered the canal. The lock-keeper at Dizy remembered us, and sold us some of his excellent limeflower honey, but he had no eggs; all the barnyard hens in France seemed out of lay at this time.

This is an interesting stretch of canal with some pretty villages, in most of which champagne is made. On the hills above the canal the vines stripe the southern slopes in rows to catch the sun. Pinot Noir, Pinot Meunier, and Chardonnay Blanc are the only permitted varieties.

Down at the water's edge blue-powdered sloe berries hung over the green water, alternating with firebright rosehips. After a cluster of residential barges at the village of Ay, the canal broadened into a lake at Mareuil-sur-Ay.

We had decided to stop at Tours to sample the previously observed Logis, which our friend at Epernay spoke highly of. We passed through the lock and tried to moor. Ha! There was, of course not enough water to get to the bank and, as the quay is on a bend as you leave the lock, we could not sit in the middle blocking the canal as we had done during the night at Dizy.

Bill can be an extraordinarily stubborn man on occasions. He had determined to eat out, not out of any disrespect for Laurel's cooking, which is excellent, but because, he said, there is no point in journeying through France if one never enjoys one of the country's greatest assets, food; and here at last was a canalside restaurant that was not closed.

With somewhat maniacal obsession the barge *Hosanna* drove herself, screws thrashing, into the mud, over and over again towards the bank. After about an hour, and some three gallons of diesel, we had attracted a small knot of curious spectators and were close enough to jump ashore over the barge's shoulders, for the fore part of the boat draws less water than the stern. The aforementioned stern still stuck out a bit, but Bill observed that it could now stay that way until after tomorrow's breakfast, speaking through clenched teeth and ornamenting the phrase with Anglo-Saxon words.

So, after having spread a few tons of mud over a large area, and levelled things out a bit, we changed and dined out. Well worth it, too. Perhaps a little expensive for a country *auberge*, which may have been why it was not very well patronised. The *tête de veau* with *sauce Gribiche* was excellent. The roast quail gave one a feeling of decadent luxury without being too filling, and we finished off with *tarte Mirabelle*.

In the morning we realised what a picturesque place we had chosen as our mooring. We could look across to the moving waters of the Marne close alongside but a bit below the canal, the morning mist wreathing and veiling the huge trees, whose few golden leaves glinted in a pale sun, framing a bridge that spanned the river.

Laurel, who had been complaining for some time that she was indistinguishable from the deck mop when it came to hair styles, except that the deck mop was cleaner, decided to risk the local hairdresser: it is after all rather an up-market village. It was a mistake. She returned with her hair cleaner, and shorter, but still a deck mop.

We stocked up with food in the village, and found that the *épicière* was not at all well. She was complaining to the six ladies at the butchery counter, and the three waiting at the checkout, that her man had left her to manage the shop *toute seule* while he went to the town. She had a *crise de foie*: a bilious attack, which, she freely admitted, was her own fault – she had been greedy.

A discussion ensued as to treatment, general agreement on the remedy was reached, and one woman was despatched to the *pharmacie* across the square to obtain it. She returned empty-handed. The chemist would have to send to town for it; it would be there at two o'clock this afternoon. Shock, horror. All clucked and tutted; that he should not have such a remedy in stock, especially as it was so good for the intoxication of the bowels in children! 'I shall,' said the *épicière*, 'close the shop until three o'clock, thus affording time for the remedy to be efficacious. It is, one knows, radical.'

All this time she was cutting veal chops, weighing sausages and parcelling up *bifteks*, occasionally dashing across to the till to check someone out. At this point one very old lady gave up the struggle. Grumbling loudly, she left the queue and shuffled out of the shop, pursued by the *épicière*'s voice extolling the virtues of patience, and a final shot, aimed through the open door at the figure disappearing across the square: 'I cannot cut myself in four, *Grandmère*!' The distant figure shrugged.

We loaded our stores back on board and left about noon. The sun had dispersed the mist, and the day was fine. Since the junction at Condé we had been back in the Canal Latéral à la Marne. A lateral canal is one running alongside a river. Where possible the canal builders will canalise the river itself, but this involves some tricky engineering to control floods. In some regions, therefore, a canal is cut close to the river, which provides the water for the canal. The river can then suit itself what level of water it keeps.

The canal had deteriorated to a state where the authorities had had to take notice. There were two stretches with 'no overtaking' signs and speed limits as low as 2 k.p.h., which is perhaps better described as slightly stopped. The commercial barges do not always keep to the speed limits; they are after all under the same economic pressure as the truck drivers.

Our next significant stop was at the old town of Châlons-sur-Marne, where we were unable to berth in the side moorings off the canal because they were all full of empty barges awaiting cargoes, so we moored ourselves in the canal just above the lock, a beautiful tree-lined *bief* (the stretch between two locks). The tall chestnuts were golden brown, and their leaves fell when gusts of wind shook them, rotating slowly as they came down into our scuppers like paper cartwheels, exciting our kitten to the chase. It took only three leaves to bury her completely.

Châlons is notable for an exquisitie church (not the cathedral) with some good glass, and for being one of the best shopping centres for a town of its size in France. Very often shops have moved out to shopping precincts on the outskirts – all very well if one is car-borne, but not so convenient for boat people. A lovely square provided good ice cream and a first-class *traiteur*, whose shop was a glutton's dream.

It was at Châlons that we ate *rattes* for dinner. Not *Rattus rattus*, nor his Norwegian cousin, and not the can of 'tinned rat' that took Laurel aback on an Atlantic crossing until she remembered that she had so labelled a tin of ratatouille, but a superior type of potato, beloved of chefs because it keeps its oval shape well when cooked, and which has become a cult in France. (Where its name comes from, heaven only knows.) We could not pass by without sampling this aristocratic and expensive spud, which sells at about two and a half times the price of the common-or-garden variety. It is a waxy vegetable, with a taste not very different from any good potato, and we decided that the whole thing was a marvellous marketing achievement by someone who had ended up with a lot of *rattes* in his warehouse.

After Châlons the locks are a bit more spread out, and one finds potential moorings beside each lock. In a well-built and maintained canal, each lock is approached via a sort of funnel-shaped narrowing of the canal. Thus one's bows are guided into the hole even if one is careless enough not to get lined up properly. The 'guides', often in a fan-shaped group of ferro-concrete piles, are known to the bargemen as *pattes d'oie*, or goose feet, and they are very welcome in strong winds. These locks were so fitted, and outside each *patte d'oie* there would generally be a short quay for barges to lie alongside when they find the locks closed after sunset.

There are also many more silos, all at this time busy.

At La Chaussée-sur-Marne the canal widens a little and there is a short grassy quay with, amazingly, well-spaced bollards. There is also a restaurant marked in all the guides, another Logis de France, that excellent chain of owner-driven inns. Yet again we lost the pleasure of a meal out, but of all the reasons this was a new one. The chef/proprietor is a keen hunter, and at weekends he is out on the slaughter, leaving his restaurant closed. This is so completely contrary to the normal out-of-season habit of opening only at weekends that we returned on board almost in a state of shock.

At Soulanges the local council has displayed commend-able enterprise and constructed a neat and well-made mooring quay for the passing *plaisancier*, with its little picnic table on a grassy bank and a clean, if elementary, WC (an ankle-splasher). It is a picturesque spot, but one is baffled as to the reasons why the council took the trouble to provide it. Having stopped, eager to patronise this gener-ous and thoughtful community with our custom, we were absolutely unable to find anything in the village on which to spend a single sou. Not a bar, even, nor shop of any description. Not a cucumber to be bought, not even for ready money. Would any readers passing Soulanges please take off their hats, or dip their ensigns in salute, to a noble and selfless village.

Just before one comes to the barge town of Vitry-le-François, there is a fuelling depot, but they are not in-terested in the *plaisance*. Their privilege; one can see them out of business before long like the next fuelling depot round the corner, whose derelict pumps stand like caryatids round the tomb of a lost enterprise.

Vitry-le-François was originally called Vitry-en-Perthois until it was razed to the ground in one of France's inter-necine wars, and was then rebuilt at the orders of François I, who gave his name to it.

At one time the canal went through the centre of the town, where there was a low-level bridge which had to open for every passing barge. As road traffic developed, this so disordered the town that something had to be done, and it was cheaper to build a canal by-pass than to try and raise the bridge and all the streets round it. The old bridge now remains closed. The original length of canal, as far as the town centre, is used as moorings for barges awaiting cargoes, for there is a *bourse* (or exchange, where work is offered) at Vitry. We promised ourselves a visit to a *bourse* a bit further south; as it turned out we ought to have visited this one. We will talk about the *bourses*, and the procedures for obtaining cargoes, in chapter seven.

We moored for the night in the old canal, now silting up. The light barges which draw almost no water are very comfortable there, but we had problems. Our neighbours, who had been waiting three weeks for a cargo, were Belgians. There exists a degree of mutuality between France, Belgium, Germany and the Netherlands regarding barge traffic. This causes some long faces among the French bargees because those of other countries have various forms of subsidy, which are not so liberal in France.

For example, in the other countries the barges are entitled to fuel at the cheapest rate, while in France barges obtain fuel at an intermediate rate. There is white diesel for general use (including yachts), and red diesel for barges, but the real cheapie is the green diesel, which only fishing boats may use. This may well be another example of the power of the fishermen's wives.

On the by-pass at Vitry there is another fuelling station, this time of a friendly disposition, together with a substantial chandler's shop. We were able to replace, at last, our broken navigation light at a price considerably cheaper than the wholesale price in England for the same model. This is a notable benefit attached to dealing with suppliers of bits and pieces for commercial craft. Often the same items are on sale as in the yacht shops, but far more cheaply.

Replenished with fresh water, fuel, rope and other bits and pieces, we continued into the Canal de la Marne à la Saône, which is the main connecting north–south link.

The railway bridge at the southern end of Vitry has very little headroom, but we were able to pass it easily as the water level had been lowered to allow an unusually high barge to get under. The canal authorities are not usually noted for co-operating with the barges; in this case we found that it was an unofficial act by a friendly lock-keeper.

The character of the canal changes again after Vitry and becomes tree-lined and attractive. All the locks are now manually operated, with no power assistance and no more of those useful *pattes d'oie* to help one in. The little quays are no longer found. Trade is thin. Most of the cargoes to be had here are agricultural products going to Rotterdam or other North Sea ports. There is also some zinc to be carried.

At Orconte, where the silo no longer sends its trade by barge, and the loading chute by the canalside creaks derelict in the wind, the bridge over the lock has buckled and dropped by about ten centimetres. It has a load limit of ten tonnes, but we watched trucks of forty tonnes cross it on their way to the silo.

The *éclusière* at Bruyère was followed as she worked by two nanny goats and a brown billy goat, stepping surefooted on the narrow lock gates with complete familiarity. At Sapignicourt lock we were counselled against using the recommended quay further on because of the presence of a tribe of *nomades* (gypsies). Another reason why a good mooring was unavailable. Gypsies are regarded in France in much the same way as they are in England – in other words, they are barely tolerated. They seem to have much the same habits, the most noticeable being their capacity to take a pretty corner of countryside and to turn it into an industrial car-breaking site and scrapyard before moving on and leaving their unwanted junk behind them. Whether they are really as dishonest as they are made out to be we have no way of knowing, but they have that reputation in every country.

We watched them fetching water in a dustbin from the

canal – an old woman in a bright blue dress, her ringlets still black as jet, helped by her man, who had magnificent white whiskers.

In the next reach we came across two rare birds; one of them, bright red, swimming under the overhanging leaves at the canal's edge, turned out to be a Coca-Cola tin. The other was a barge with an all-English crew. She was called *Perche* and was registered in Belgium. The crew told us that foreigners were not allowed to operate French barges, but that Belgium was much more tolerant and, provided that the barge was legally registered in Belgium, they could trade with an English crew. Not a brilliant living, they said, but with hard work and long hours one could achieve a reasonable and independent existence, which they enjoyed.

Had we heard news of a possible *arrêt de navigation* further downstream? No, we hadn't. 'Keep your ears to the ground,' said *Perche*. All this conversation was conducted by radio, on VHF Channel 10, which is reserved for the bargees' use; it was virtually the first time we had spoken on it instead of just listening to them talk, and understanding about one word in ten.

This canal, from the Marne to the Saône, has a number of small towns along its route, and the next was St-Dizier, called after St Desiderius. It is famous for its defence, with 2500 men, against the 100,000-strong army of Charles Quint in the sixteenth century. (These numbers are given in the French history books; if true then it really was a great achievement.) The King of France, who was somewhere else at the time, was understandably very enthusiastic about this victory, crying: '*Allez, braves gars*'. The latter two words were later corrupted into *bragards*, which is how the inhabitants are referred to nowadays. It was further corrupted and underwent a different shade of meaning in English.

St-Dizier is distinguished from a boatman's point of view

for a good quay (only slightly spoilt by the ever-present dog turds), with a wine *cave* alongside where one can top up with draught wine and cheapish bottled products. There is also a garden centre on the quay, where we watched a barge restocking with the ubiquitous geraniums. We were passing up on horticulture at the time, making do with a pot of parsley and sweeping up the daily sackful of fallen leaves in the scuppers if we felt like it, egged on by Tassie, who loved gardening.

St-Dizier has a lovely garden in the Place Winston Churchill, a street-name surprisingly uncommon in France when one considers the enormous service that he rendered that country. (Come to think of it, there aren't that many in England either; perhaps it is necessary to be shot to become a hero.)

There are, however, numerous streets in France named after Wilson (Woodrow of that ilk, definitely not Harold). Compared with England, streets in France are more often named after people, and perhaps the two most common names are Gambetta and Jean Jaurès, two people almost completely unknown on our side of the Channel.

Gambetta, known to us affectionately in garbled translation as 'the little shrimp', was a formidable statesman of the mid-nineteenth century who achieved fame for the gallant way in which he organised resistance to the German invasion of 1870, and by his implacable opposition to a proposal to submit France to the temporal power of the Pope. Still, it is difficult to see why he is quite so popular as a street-name.

Jean Jaurès is more understandable. He was a socialist Deputy in the period preceding the First World War, a newspaper proprietor of outspoken views, and a practical pacifist. He tried hard to organise the workers of France, Britain and Germany to refuse to fight and, by so doing, prevent the imminent war. He was gunned down in early August 1914 in highly suspicious circumstances that seemed to involve the ruling 'establishment' in France. As with all martyrs in a good cause, however mistaken they may have been, he became a national hero, though again he is little known in our island. It would be difficult to find a French town without a *rue*, *avenue*, *boulevard* or *place* named after him.

We were now approaching an even prettier part of the Marne valley. Its sides were becoming steeper, the woods denser, and the surface of the canal was thick with falling leaves, more spinning down with every gust of wind: gilt lances from willows, gold pennies from poplars, and yellow hearts from bindweed.

The lock cottages were still bright with flowers – asters, chrysanthemums and dahlias – planted in tubs, barrels and pots. The cottages are all the same: the narrow end faces the canal, and steps lead up to a shuttered front door with a *marquise* (a little glass canopy) over it. Above the doorway is a window, also with shutters. This window is usually open, and in the morning the bedding hangs over the sill, with the lace curtains blowing in the wind. There are more windows at the sides of the building. The shutters are painted pale green, and the cottage is rendered, with brick corners.

Given this basic cottage, which he rents from the canal company, the fancy of the *éclusier* and his family then takes over, finding expression in the flowers and shrubs, and the *basse-cour* full of chickens, ducks and geese. Some have pigeons, rabbits, goats or sheep; some have fences round the door to keep the children in or the dog out. Outside the door hangs a lifebelt; there may also be caged birds, geraniums and washing.

There will be at least one dog, usually large and noisy, and quite often cats as well. Less often one finds plaster storks, gnomes, model windmills and wheelbarrow planters. All lock cottages have on the end wall a bright light to illumine the lock at night, and keep somewhere handy a small collection of interesting tools on long poles.

These may be used to take your line if you are at the bottom of a deep lock and the *éclusier* is feeling friendly, or to free the lock gates of obstructions, occasionally gruesome.

The canal has some narrow stretches here with attractive villages. The by-roads cross it on rickety swing- or lift-bridges, some of them operated by very old and equally rickety pensioners. All these opening bridges are manual, and they all take a long time and a lot of creaking to work. There is not much to be bought in these villages; many do not even have a bakery, the bread arriving in the back of a small van once a day. Opportunities for eating out are rare.

We were hoping for such an opportunity at Rachecourt-sur-Marne. We made it a night's stop and headed for the restaurant. The quay was really not a bad mooring, which made a change. It had rained heavily, and the long grass through which we ploughed in the gathering dark felt very wet. Little stones seemed to be getting into our shoes – we try to avoid going to restaurants in our seaboots. The restaurant, which had been deleted from the latest Michelin Guide, was OPEN! We paused in the entrance to shake the stones out of our shoes and were disconcerted to find that they were dozens of little white snails. No wonder they felt a bit wet and crunchy. After a couple of restorative drinks in a large bar empty save for two children doing their home-work, we went into the restaurant, where we seemed to be alone but for a handful of travelling salesmen, and were persuaded to have the *specialité de la maison*. We'd have done better with *steack-frites*, which is what the salesmen were having, as the waitress was archly secretive about the 'special'.

'What is it?' we asked as she put it before us. She replied: '*Devinez!*'

Nowhere else have we ever been asked to guess what we were eating, and nowhere else would we have failed. We think we know why Mr Bibendum deleted this one from his Guide. We think we agree with him. We have deleted it from our notebook too, but somehow we cannot forget the snails.

We returned on board for our coffee and a Calvados, pausing to remove another crop of snails from our socks, and regretting that they were much too small to eat with *beurre de gascogne*, and anyway rather squashed.

When we left next morning we could see the snails bunched like blossoms on a branch, clustered high on the grass stalks in order to keep their feet dry, which explained how they had dropped into our shoes as we brushed past.

We passed through the Chevillon lock at Rachecourt after a night of heavy rain. 'It's getting cold,' we said to the lock-keeper. 'Oh yes,' he replied, nodding towards the huge woodpile stacked there. 'Soon it will snow, and the ice will come.'

The canal was covered with fallen leaves, and the water swept them thickly into the eddies and *diversoirs*. The mornings were chill and misty, and only a frill of leaves remained at the very top of the poplars.

It seemed that we had better get on south as fast as possible. So we headed for Joinville, which was one of our mail-drops, but were quite unable to stop there because the grassy bank in the town centre was not only bollardless, but also inaccessible, so we had to stop much further on and trek back to the post office. Smaller craft can often moor at a grassy bank, and secure themselves to stakes which they hammer into the soil. Heavier boats like *Hosanna* pull out the stakes.

The French Post Office workers were *en grève*. France was undergoing a miniature version of our winter of discontent at that time. The news was full of strike reports, and interviews with trade-union leaders who seemed extraordinarily like the British ones and spoke in similar clichés. Bill is inclined to get rather blimpish on the subject of strikes by public servants. He has always maintained that as the Civil Service is very seldom civil, it should at least provide a service. It will astonish nobody that our mail had failed to arrive.

It never did catch up with us, though we made great efforts to go back for it, or have it sent on. Messages were passed on by willing lock-keepers regarding where to send it next, and we filled in numerous forms giving la Poste awesome powers to redirect our mail to what appears, in the end, to have been a Black Hole. All this, had we known, was but a small inconvenience compared with the week that followed.

6

The arrest of the navigation

'Mai s'avien d'èstre, Dequé deviendren tant d'ome e d'ome que vivon dou trahin de la ribeiro . . .'

. . . .

'If that were to be, what will become of so many men who live by the work of the river . . .'

There was a long delay at the lock at Gudmont, which was closed against us. There was no one about, no acknowledging wave to say: 'I've seen you, the lock will open soon.' Nor was there anywhere to moor, so after a while we put *Hosanna*'s bows into the bank and Laurel scrambled ashore to see what was happening. The lock-cottage door stood open, as did the upstairs window. No dog barked, no one came hurrying out of the shed. Calling *'Ma-Dame!'* brought no response.

A glance through the open door into the kitchen lighted on the remains of lunch, a cat asleep, the clock ticking – all eerily abandoned, like the *Marie Celeste*, empty of human beings. The only other visible building being the crossing-keeper's cottage, Laurel walked over the railway line and ran the *éclusière* to earth, gossiping in the kitchen with the crossing-keeper. She started like a guilty thing upon a fearful summons, leapt from her chair as if stung, and was back at the lock and opening the gates before Laurel's leisurely tread got her back to the boat, so for

once she watched *Hosanna* enter the lock with a tourist's
eye.

It is a very rare event for a lock-keeper to be absent; for
one thing their telephone is always ringing with news of
events and catastrophes up and down the canal. The *éclusière* gave us no news as we passed through. It would have
been better for us if she had done. Three locks later, at
Froncles, where we had thought to spend the night, rolling
the jolly little name round our tongue, we came to the first
hint of the *arrêt* that *Perche* had mentioned. It was half past
four, and we were ready to stop. We asked the *éclusière*'s
advice on a mooring.

'It is sad,' she regretted, 'but the restaurant at Vouécourt
will be closed and the mooring at Froncles is not convenient. Do you not know that the navigation is to be
arrested? You should concern yourselves with finding a
good place to stop.'

'We have heard rumours, but no one has told us when or
where.'

'We also, they tell us nothing. But they will know further
on at Relancourt, which is a *poste de contrôle*.

'Very well, we will direct ourselves to Relancourt, but
where is the *arrestation* to be?'

She was vague. She waved onwards, saying *'plus loin'*,
which we interpreted as 'off my patch'.

Time was getting on, but we made good progress and
arrived at the lock at Granvaux at ten past six. We should
reach Relancourt, which was still some distance on, near
Chaumont, tomorrow. It was deep dusk and the lock had
already closed for the night. A barge waited, moored up
ahead of us.

As we made towards the right bank, which had all the
appearance of an overgrown, bollardless heap of earth, two
figures advanced through the gathering gloom, searched
for the bollards, which actually existed, and helped us moor
to them. They were authority. They had information.

Thus we came up with the first hard news of the *arrêt de navigation*. The canal engineers had been planning a *chômage* to carry out repairs at various places in the summit pound near Langres, including the vault of the tunnel of Balèsmes. Events had been precipitated when a lock-gate hinge had failed on the south side of the tunnel. No further traffic would be able to come northbound, and the southbound traffic could descend as far as the affected lock, provided it was done as soon as possible.

'OK,' we said in our best French, 'we are not pressed, we will stay here in some pretty little spot, where there is a grocer and a restaurant. We will await the end of the *chômage*.'

Well, no. Alarm and consternation. The idea of a barge not being in a hurry had not been programmed into their problem-solving mode.

'But it is a matter of urgency to recommence the navigation at a good hour in the morning. The locks will reopen at six o'clock.'

'But at such an hour it is still dark, and we do not have the lights for night navigation. They are not, after all, obligatory.'

'But there are barges behind you who must continue.'

'OK, we will go on if it is not dangerous to do so.'

'At six o'clock, then,' and the ambassadors of the *navigation* left, stumbling along the grassy towpath, clearly troubled, talking animatedly. We could still see their arms moving as they regained the road and their car, some two hundred metres away.

At midnight our sleep was disturbed by threshing and swirling, and the throb of a motor as another barge moored close by.

We were up at five forty-five. The barge that had been berthed ahead of us was already going through the lock. It was pitch dark and very foggy, visibility about one hundred metres.

A wavering lamp along the path illumined the lock-keeper. 'Why were we not preparing to follow? There is haste.'

We explained that we would prefer to wait till daylight, and were happy to allow the barge that had berthed at midnight to pass ahead of us. With this everyone was content, and we finally locked through at six forty-five, in a grey and foggy dawn, following the barge called *OK*.

A pilot system was in operation to clear the barges through as fast as possible. This meant that one man was with us all day, going ahead to fill the next lock and open the gates, since there was no traffic *amont* (coming up) to set the lock the other way. The name of our pilot was Daniel, and during the day he began to understand our problems, and we to understand his. The stretch of canal was to be completely emptied for urgent work to be done, and all barges had to be out of it before the drainage began.

We explained that we had no searchlight as the commercial barges had, and travelling after dark was therefore imprudent, not to say dangerous.

We also explained that we had no bread.

We heard him telephoning his superior and outlining our difficulties. He came back with the news that we would be able to stop at Rolampont for as long as we liked, 'three weeks even!' he said grinning. We would be blocked there for eight days, the duration of the *chômage*. We didn't mind. We were much relieved.

With the realisation that we had no bread, a deprivation which would touch the heart of any Frenchman, our pilot paused at the next lock while he took us to the baker's in his little orange van. All of us had lunch as we worked, swigging down a beer and eating a half *baguette* with hands none too clean from the constant handling of muddy ropes.

We said goodbye to our pilot at Foulain lock at half past five, giving him a bottle of good wine as a small token of our esteem. We were warned to be up early on the morrow to follow the same procedure. But tomorrow, said Daniel, it will be the *dames à Mobylette*.

After more than twelve hours' travelling, and a record (for us) of seventeen locks in one day, we were very tired. We went on a short way and moored with difficulty in a broad curve at the small town of Foulain, where restaurants had been promised faithfully by all the lock-keepers, and stumbled off in the dark to look for them.

We walked up the hill over the level-crossing into the little town. It is not a very prepossessing spot, even though its riverside is attractive, and it owes the facility of restaurants to the busy trunk road that splits it in two, making contact between the two parts as difficult as, but much more dangerous than, crossing the Channel.

We expect the reader is ahead of us. It was a Saturday, the Buffet de la Gare was having its *fermeture hebdomadaire*, or weekly closing day, and Le Châlet was reserved for a wedding banquet, and didn't want to know two tired and not very elegant amateur *mariniers*.

Further back, near where the lane from the canal joined the main road, we had noticed (but scorned) the Café du Centre, *snack à toute heure*. Laurel, dragging behind, observed that she was on strike as far as the kitchen was concerned, and it would have to be here or nothing. We went in.

There was quite a crowd. There was a juke-box. There were the usual children doing their homework in front of the television set. There was a notice saying *'paniers accepté'*, meaning that one could bring one's own food if preferred. We had a drink and debated the situation. Though snacks were the advertised fare, Madame was quite prepared to cook us a simple meal. She did. It was not *haute cuisine*, but it was cheap and welcome, the lady was friendly and hospitable, and we were very grateful. Back at the barge, we fell into bed.

We started to get under way again at dawn, as we had been asked to do, and while making preparations we waved on the last barge in the string, *Torne*, allowing her to go ahead of us. Now we were tail-end Charlie – carrying the red lantern, as the French say.

Again we were allotted a 'pilot' to work us through the locks, a series of ladies this time – the *dames à Mobylette*, with their neat little motorised bikes. A subtle distinction: the men, it seemed, got the orange vans, but the ladies were resplendent in new orange Anoraks, and glad of them since the day was cold.

We found we were continually menacing the backside of *Torne*, the barge ahead of us, which was doing likewise to *OK* ahead of her. It is always difficult following a fully laden barge. They are very deep in the water, and the canal was silted and shallow. Even with our engine at its idling speed we were continually catching up.

Navigation was therefore tenser than usual, and a lot

more tiring owing to the higher standard of concentration required.

At Prées, after our lady pilot had made a telephone call, we heard the bad news. Yet again plans had been changed, or had been previously misunderstood. We could not stop at Rolampont, where we had arranged for our anxiously awaited mail to be sent on from Joinville. That *bief* was to be drained as well.

The cry was now '*Á La Une!*' or ever onward to lock number 1.

It was at this point that Bill's stubbornness came once more to the surface. He called heaven to witness, politely but firmly, that not one metre more would he navigate until someone in authority told us just what the devil was going on, and where. There was alarm and consternation – they could not pull the plug with a barge still in the lock.

More telephoning, and then a short and slightly anxious wait while *Torne* disappeared slowly out of sight. No matter; we could soon catch her up if we decided to end our mini strike.

Evidence was soon forthcoming that we might be upsetting the plans of the Powers as much as they had upset ours. A larger car than the one at Granvaux arrived. The heavy mob. Not wearing gumboots this time: leather shoes, muddy by the time he reached us. Smart suit of country cut. Tie rather like the Old Etonian, stripe a little too broad.

First the pleasantries. The French are always extremely polite (except after causing a motor accident); indeed it can be said they are an example to us British. Then the business. Patient voice trying hard to speak English, getting a little lost pronouncing words like 'vault'. Very relieved when we invited him to speak French; he had assumed that the failure of the others to put over their message was due to a language problem.

They were not only going to repair the lock gate, they were going to drain about thirty kilometres of canal while they were about it and take the opportunity of doing other important repairs too. They were also going to repair the stonework in the vault of the tunnel. Notices had been affixed in the *bourses*, surely we were *au fait*.

'But the *plaisanciers* do not go into the *bourses*.'

He paused to consider this news. It dawned on him that we were right.

'Also,' Bill continued, 'every lock-keeper tells us a different story – even your own people are not *au fait*.'

We now started a series of full and frank exchanges. We made it clear that we were not wishing to be difficult, but we had to make some plans, we had to buy food. We could not navigate safely by night, we had our mail to pick up, and it was urgent. The position was almost insupportable, but could be improved if we knew what was taking place.

He was only responsible as far as *La Une*, lock number 1, he explained. Proper arrangements had been made after that; we would be told in plenty of time. He would send a car for our mail. There was great urgency to start the work.

'Could we go back?' we asked.

No, we could not go back because we could not turn round. We could not go on to the turning basin just short of the tunnel and turn there, because their programme was to use all available time, and as we were the last barge to enter the affected zone, they could not afford to wait while we got as far as the tunnel and then returned. The plan was to drain each *bief* as soon as we had been let out of it into the next one. Would we please therefore go on with all haste. It was very important – the cost of a delay would be very great.

It was a lost argument. We didn't want to be the party-poopers at the Navigation Ball. We apologised for deranging him; he apologised because correct information had not reached us earlier. He departed in his car, and we got under way once more.

We caught up the barge ahead of us within an hour. We

were forced once again to plod along at a speed that was below our engine idling speed.

One cannot just stop, because one drifts all over the place, so we had to keep moving. It was a case of neutral – slow ahead – neutral, every few minutes. At the end of each short driving spell we were dangerously close to touching the barge ahead of us in a place where gentlemen simply do not touch other gentlemen.

The barge ahead observed our problem. Our attempts to listen to the bargees' conversations had been thwarted by the *argot* they spoke, which we found very difficult to pick up. In addition, almost all non-military users of R/T have very poor radio discipline, and there is much err-ing, and overlapping, and long pauses.

We had conversed with *Perche*, which had an English crew, and with one or two English-speaking Dutch crews, but had not tried to communicate with French or Belgians until now, so were startled to hear: "*Osanna*, ici *Torne*,' coming over our radio.

They called us in rough French. They explained that they could not go any faster because the canal water level was lower than normal, and they were very deeply laden, and were having to plough their way through the mud on the bottom using full engine power. They said too that the boat ahead of them had the same problem, and the crew was angry because it was wasting a lot of fuel. We replied that we were not too pleased either, we simply could not go slowly enough, and it was not good for either our tempers or our engine. We had to keep close up because the water was being drained out of the canal as we passed each lock.

There was some rapid French on the radio, which clearly did not concern us, or in which we were not invited to join. We heard from one boat we had not yet met, '*J'en ai marre*', a remark that received grunts of agreement elsewhere; in truth, everyone was fed up.

As *Hosanna* passed each lock, gangs of men would drive

off in orange vans to open sluices and drain the *biefs* behind us, and we imagined the lock-keepers saying 'Pouf!' and putting their feet up for a few days.

For us there was a definite feeling of pulling up the drawbridge, or of cutting off one's past. There was only now a future tense as we were swept along by a remorseless force, knowing how the spider must feel when it is caught up in the housewife's broom; past Humes, number 5, Jorquenay, number 4, a very pretty village with a centre-pivoting swing-bridge.

Darkness fell, our second pilot went off-duty, and a third took over. Still we went on, inexorably down to number 1, *La Une*.

Emergency lights had been rigged at each lock, to augment the usual lock-cottage lamp, which helped a great deal. We were socially isolated, with no one to talk to and no rest – cut off in a dark, misty world, apart from the lights of an occasional hamlet; for it is a sparsely populated area. Now and then, when the mist cleared a bit, one could make out the bright lights of the city of Langres high on its fortified hilltop some miles off to the westward. We thought of the snail soup we had more than once enjoyed in the Auberge Jeanne d'Arc there, and moaned a little. This, we thought, was not *plaisance*.

Past Moulin Rouge, number 3, and Moulin Chapeau, number 2. There was little talk on the radio; only a Belgian lady from the barge *Regina Coeli* chattering to some unseen and largely unheard compatriot. We raised a chuckle, speculating on who the 'Queen of Heaven' might be talking to. Evidently the King of Heaven her shepherd wasn't; we could not follow a lot of what she said, but the tone was undeniably disaffected.

Then at last we were through *La Une*, and our tired pilot disappeared rather suddenly. We were right at the top, 340 metres high, in the summit pound on the Plateau de Langres, the watershed between the rivers that go north and west to the North Sea and the Atlantic, and those that go south to the Mediterranean.

Hosanna in the highest, you could say, and from here on it ought to be all downhill. But the trumpets, it seemed, were not sounding for us on the other side.

La Une, Batailles number 1, the first lock on the Marne (if you are going north) the last if you are going south. La Une, where the gates of paradise should have opened for us, where we had hoped for information, hot coffee, bollards every two metres, a deep grassy quay with no dog turds and, if no restaurant, at least a supper in the warm and a long night's sleep.

There was nobody about, and *Torne*'s searchlight, way ahead of us, flickered over the steep wooded banks that blocked the moonlight and afforded no mooring, eclipsing the torch that Laurel was shining on the canal banks.

We should have realised that once we were past La Une we were committed to the tunnel. We closed up a little on *Torne*, taking advantage of her searchlight, and an hour after leaving La Une entered the tunnel of Balèsmes, which the barge people call the Souterrain de Heuilley-Cotton.

To our surprise, *Torne* started to accelerate a little: the water was clearly deeper. There might have been another reason. It was as well that we did not know then that slow-travelling barges were sometimes attacked and robbed in this dark tunnel, nearly five kilometres long.

We were tired, and concentrating was not easy, as *Torne* led us with her kindly light. As we felt our way through, so slowly that our motor was hunting at one 'rrrumph' per second, we scraped along the walkway to our right, twice committing the serious crime of losing one of our tyres. Attempting to retrieve them from the black water behind us did not seem a good idea. We bequeathed them to the engineers who would be working there on the morrow, presumably with nice bright lights.

As if all this were not enough, the smell of frying onions

wafted back along the tunnel from *Torne,* assailing our nostrils like the scent of Paradise Lost, and driving us mad. Lunch, snatched between locks and not lavish, now seemed like yesterday. Both of us were needed to keep a lookout, and we did not dare even dash downstairs to fetch a beer.

We whiled away the time by dreaming up a Water Feast. It should start, we thought, with watercress soup, and after some discussion we thought trout, or a water bird such as duck, was a more practical main course than such mediaeval dishes as heron pudding or roast swan. The vegetables would be snow peas and water chestnuts. For dessert we argued between Floating Island and Sussex Pond pudding, with a nod in the direction of water ices, but then remembered the 'feet-in-the-water-blackberries' that used to clothe the rond in Norfolk, growing fat and luscious on the moist river bank. Suddenly we were back in our teens, two determined youngsters in a dinghy, one of us rowing into the brambles, the other gathering the fruit and getting wet feet.

We decided, forty years on and in the tunnel of Balèsmes, that watermelon stuffed with these blackberries would continue our imaginary banquet in the right spirit. We would drink our wine (Lacrima Cristi, perhaps, or Entre Deux Mers?) from Waterford glass, eat our cheese with water biscuits, nibble a few river plums, and round the feast off with Eau-de-Vie. The time passed remarkably quickly.

After about an hour we saw *Torne*'s searchlight pick out the tunnel's end and pass through it. There seemed to be some moonlight out there. We emerged, however, into steep thick woods, which cut the light down. We followed *Torne* for another three-quarters of an hour, when her searchlight began to swing first to one bank and then the other, and then to settle on a large tree, like an anxious dog.

Torne had decided to stop.

The moon shining through the mist revealed that all the

barges had stopped for the night; for here, about two kilometres past the southern portal of the tunnel, there were places to moor. The barges had stopped at all angles to the line of advance, as they had each moored with only one rope to the bank and had let their sterns drift willy-nilly. The crew of *Torne* were having difficulty getting a line ashore as, being deep laden, the barge could not approach the bank, so they lowered a rubber dinghy. We managed to get our bows into the bank, and Bill jumped out to make fast to a small tree.

Once safe, we could relax our tired eyes and aching muscles, pour a drink, start the generator and cook ourselves some supper. Then we sat in the wheelhouse with a *digestif* and listened to the conversation on Channel Ten, for mooring (and perhaps a little relaxing alcohol) had loosened many tongues with much to complain about.

It should be understood that none of these *bateliers* had been able to make a phone call for two days now, and they were to be held up for an indefinite period. They needed to get urgent messages to the consignees who were expecting the freight that there was a delay. The barge skippers had just discovered that here in the village of Heuilley-Cotton, where we had been brought to a halt, there was not even a telephone, let alone a bakery. Their wrath was expressed in forceful and vivid terms. One had made a whole salad of it, they said. The verb *caffouiller* (to make a horlicks of) was heard.

One can only call it a lack of consideration on the part of *la navigation*; it is possible they had no option, but that was clearly not the opinion among the barges.

We slept soundly, morale restored by warmth, rest and food, and the feeling that we were not alone.

Just before first light there was an earnest conversation on the bank to which we had not been formally invited, but which Bill attended as observer. There a misty figure, authority in a dark raincoat, told a murmuring gathering that they were moving us on to a village called Villegusien, where there would be *tout ce qu'il faut*.

'There is a baker, and a butcher also. There is a post office and telephone. There is even the *établissement* of the family Dupont.'

There were a few rapid questions that Bill's French could not cope with, but which were answered with 'Four or five days'; this seemed to be the expected duration of the stoppage.

The shop stewards seemed to be satisfied with the management's proposals, and negotiations were concluded with a return to work. Everyone went back to their boats, and the long process of untangling seven barges from wherever circumstances and the wind had blown them overnight began, with a great deal of crashing about in the reeds or the dense scrub according to which side of the canal had been chosen for the mooring rope.

It took two hours, and since we were the last it enabled us to have a good cooked breakfast as we did not know when we would get another bite to eat. Finally, as tail-enders once more, we followed *Torne* onwards. Barges are very strict about maintaining precedence.

We were now on the Saône side of the watershed, and descending fast through a chain of very deep locks. A different technique was required for mooring in them. Instead of having to throw up a rope to the top (where with luck the lock-keeper would drop it over the bollard for you) and tend it as the boat rose, one now had to take care not to leave one's rope behind as one descended. Ladders are not Laurel's strong point, so she quickly found a longer rope, fastened it firmly to *Hosanna*'s mooring bitts, and looped it round the bollard at the lockside, feeding the line out as we descended. It was then a simple matter to flip the rope off the top bollard if you were clever, or at worst pull the loose end back to the boat.

This was just beginning to work rather well when we

were held up at the entrance to number 7 lock, where we had to wait while the crew of *Torne* landed the motor car she carried on deck. The lock-keeper stopped emptying the lock chamber when the car's wheels were exactly level with the edge of the lock, and they drove it ashore over two planks, with practised ease.

We didn't fume all that much, really, but it was very frustrating, especially when they did not even then move on, but had some sort of social gathering and opened a bottle of wine. When we eventually entered the lock an amiable lock-keeper explained that it would have been no use *Torne* leaving, for the next lock was only a hundred metres further on and the barge ahead had not yet entered it. It takes about twenty-five minutes to work a lock, and another ten to prepare it to admit the boat behind, which is why it saves a lot of time if alternate boats are going in opposite directions. That was not the case here, since we were all southbound.

It is quite normal for there to be a traffic jam when a line of boats comes to a few short *biefs* with closely spaced locks. We found that in this case, as well as having a bit of coffee-housing together, the lock-keeper had persuaded *Torne* to adopt one of their kittens. We had to wait here too, so we went to see the kittens, but we were already over-stocked with our one little terror, who enjoyed locks and had taken to rushing up and down the cabin roof, digging her claws into the canvas that covered our wood store, in a series of knight's moves, semi-sideways jumps with arched back and legs as stiff as pit props, tail bristling like a bottlebrush and growling at everyone on shore.

We moved on, and soon caught up the slow-moving *Torne*, who reminded us of Gray's ploughman plodding his weary way.

Then we were through the last lock of the chain, there were no more events, and at lunchtime we arrived at Villegusien to find the barges ahead of us straddling the canal as if they had been swept up there by a tidal wave.

We put our nose into the bank in line with *Torne's* bows, just behind *Regina Coeli*. The King of Heaven, in a woolly cap, helped us to make fast to two poplar trees across the towpath. This left our stern sticking out some three metres, where it almost touched *Torne*, whose crew requested permission to walk across us to the shore, *Torne* being too deeply laden to get near it. Between us we blocked the canal, but no one else would be coming past for a long time.

The deeply laden barges were mostly stuck in the middle of the canal, for the level had been lowered a little, and we now saw why they all carried five-metre gangways to reach the shore. It is one of the sights of canal life to see crews (and dogs) of all ages nipping along these long, narrow planks which spring under foot unless you are careful to walk at a pace that doesn't coincide with the natural harmonic of the plank, a technique the *mariniers* have by instinct.

We were nine barges in all. Two Belgians, *Taxi* and *Beethe*, had been there already loading agricultural fertiliser at the silo quay, which they completely filled. Far ahead were two Dutchmen, *Varia* and *Henjo*; three French barges, *Barracuda*, *OK* and of course *Torne*; another Belgian, the *Regina Coeli*, whose lady was still rambling on the radio in her lace-curtained wheelhouse; and ourselves bringing up the rear for the British Empire.

It had taken us a deal of hard work to get ourselves settled down for a stop, so we rested during the afternoon. After tea we went ashore for a look round and found a rather sleepy village, with few signs of any active life. The village store was open, and we bought new batteries for our torches, which had all expired during the last few hectic days, and made some inquiries.

'Is the restaurant open?' we asked.

'Of course', they said, then doubt clearly assailed them and there was a debate which entailed Grand-mère being called in from the back room, as well as son-in-law from next door. Finally, yes, they thought it was still open.

We walked there, crossing over the busy little river Vingeanne, and passing the war memorial. This last was a modest obelisk, a few feet high. To improve the surroundings, yew trees had been planted on either side; their roots must have grown right down into the Vingeanne, because they had grown enormously and, although they were neatly clipped into cubes, the little obelisk was crowded and dwarfed to nothing by these gigantic trees.

Over the level-crossing, we saw the lights of the restaurant beckoning, and our spirits rose. We found the owner's wife feeding her baby beside a small bar with a pram alongside it. Desolated, but the restaurant was closed – there were not enough people at this season. However, one can serve you a drink? One certainly could. We had a drink before the dark of the long walk back. The restaurant opened only in the high summer when campers came to enjoy the lake-side. Most of the campers are Dutch; it is interesting how resorts become the reserve of one particular nation, whose people come back year after year.

It really was a long way back. Laurel, who normally finds three hundred yards the equivalent of an army route march, had been spurred on by the thought that she would not have to cook dinner. With no such star to follow on the return she flagged, and we only just made the Bar du Lac in the village, an unpretentious country café, and a scene of some liveliness. A good few of the clientele had arrived on tractors, which were parked outside. The principal source of livelihood for the village was obvious. Nearly all these farmworkers had thoughtfully left their muddy boots outside, and were wearing carpet slippers of the kind known in France as *charentaises*.

We discussed the *arrêt de navigation*, and the fact that there were nine barges stranded in the village, which astounded the company as no news of it had as yet penetrated, so cut off from the canal life is the whole community. They talked of a similar occasion a few years back when the canal had frozen solid for weeks, and several barges had been stuck in the ice throughout. 'But yes, Madame et Monsieur, it freezes hard in winter.' Not surprising – the plateau is 400 metres above sea level and a long way from the warming effect of the gulf stream.

Fortified and restored by the cheery good humour of the Bar du Lac, we made our way back to *Hosanna* for a simple supper and a tired night's sleep.

What appeared to be another meeting of the shop stewards gathered on the quay at about seven-thirty next morning. The tone of the meeting was undoubtedly grumpy. It was definitely a men-only meeting, and the women were not represented; the only female voice to be heard was the Queen of Heaven's Belgian accent still talking on Channel Ten to someone far away and indecipherable. For all we knew she had been doing so all night.

Bill was despatched to the meeting as observer once more, and he arrived on the bank at the same time as authority, for whom the company opened up, enabling both of them to enter the circle. This was M. Bernard, a stocky man of about forty, with the open, friendly face of an East Anglian farmer, cheeks red from working outdoors, jerseyed under a tweed sports coat, and corduroy trousers tucked into his gum-boots. He had an engaging smile; he would need it.

He turned out to be the divisional engineer in charge of the whole operation, and he was aware that he had trouble to cope with.

Start off with a joke at his own expense; that's a good technique, Bill thought, as the engineer confessed to having made a bad bloomer. The baker was closed for his annual holiday.

This was no joke. There was a stunned silence. Worse disaster could not have befallen the community. Lightning and thunder, storms, heavy rain, broken legs, bleeding wounds, even perhaps another invasion by Germany; these things can be shrugged off by the French. They can mobilise forces to counter them, and if they fail then what cannot be cured is to be endured. But to be deprived of daily bread; there is no remedy, this is the stuff that breeds revolution.

The engineer's back may have been against the wall, figuratively speaking, but he was a man of flexibility and enterprise. He could not win, but he had not yet lost. The barges could not travel on, and all agreed that there was no village with a baker anywhere near the canal. If the barges cannot go to the bread, the bread must come to the barges. A lock-keeper's wife was appointed to drive to the nearest town to bring us each day our daily bread.

Bill remembered that the restaurant was also closed, but

as this didn't seem to concern the rest of the group he did not propose that *la navigation* should provide us with meals on wheels. Once upon a time the barge folk used to foregather for revels in canalside inns which still exist under the evocative name of Bar de la Marine, but that was in the days when there were ten times as many barges, and there was not the haste and deadly earnest that there is now, when barge crews travel from pre-dawn to post-twilight every day that they are laden. In times past these bars would have had dances and *kermesses*, which would have provided the social life that led to the intermarriage of barge son with barge daughter. Nowadays, these gatherings are rare; sons and daughters marry out of the *batellerie*, and more barges drop out of the trade.

Now they keep in contact largely through the bargees' newspapers, such as *La Vie Batelière*, edited by a priest in Vitry-le-François. It gives the latest news among the community, boats wanted or for sale, marriages and deaths, reports of such barge fêtes as still take place, and birth notices: 'I am called Camille and I have the joy to announce my appearance in the world on 11 March. My grandfather and grandmother on the *Xavier* are very happy.'

We had brought a load of joinery timber from Great Yarmouth under a canvas cover on the deck of *Hosanna*, and during the day Bill occupied himself setting up a bench on the bank and cutting out pieces to make the carcasses of twelve drawers for the galley, an activity which excited the curiosity of the barge folk, who at intervals gathered round to watch, criticise and make suggestions. Barge folk are in the do-it-yourself culture too, but *plaisanciers* are conventionally regarded as rich enough to pay others to do these things.

Having established that we existed at a similar level of domestic economics, the way was open for further fraternisation. The news that Madame *Hosanna* was demanding thirty-one drawers brought the first real smile of the day to the bargees' faces. Their eyes rolled and they made bets as to whether she would get them.

Laurel went on a shopping expedition in the forenoon, to the remarkable Ets. Dupont. The building, standing on a corner, did not look at all remarkable. It was furnished with the usual shuttered windows, and a bicycle rack stood outside. A very modest modern shop window, which three giant packets of detergent almost completely filled, gave no promise of the range of goods inside.

It would be hard to think of an item that could not be found in this shop at the right season – the problem would be looking for it. Mme Dupont's filing system was imaginative to say the least, resulting in strange cohabitations. Asking for the torch batteries the day before, we had been directed to a corner full of fishing tackle.

Groceries went with baby clothes, haberdashery with postcards; there were brushes (house, paint and flue) with exercise books and stationery; knitting wool went with boxes and boxes of *charentaise* slippers; there were tools and utensils and garden equipment, oil, turps, paraffin, paint, gifts and souvenirs, plates, cutlery, casseroles and saucepans. An upstairs gallery contained anything classified as a Large Object, whatever its use: huge stoneware jars, carboys in wicker baskets, gas fires, airing-racks, chicken huts, greenhouse-warmers, hose reels, tilley lamps, oil stoves, and the spares to go with them – all in a space equal to one floor of a modest house. It reminded Laurel very much of *Hosanna*, with its present lack of drawers and cupboards. We didn't have a chicken house, but we did have a greenhouse-warmer; it had kept the engine-room from freezing on our winter journey from Holland.

Skirting a rail full of huntsman's jackets, and nearly falling over a pile of thigh boots, Laurel approached the counter and paused behind a lady with neat grey curls and a smart skirt and cardigan. Her voice was instantly recognisable as that of the Belgian Queen of Heaven, recalling with Mme Ets. Dupont the terrible winter when she had been caught in the ice for three weeks, here at Villegusien, where it freezes hard in winter. Was that twelve years ago, they asked each other, or thirteen, using as a yardstick the year of the demise of the regretted Monsieur Ets. Dupont.

Barges can be trapped, they explained, either by ice several feet thick in the canal itself, or by frozen lock gates. A good lock-keeper knew how to keep the gates free of ice even at the worst of times, those less skilful or not so strong might allow them to freeze solid.

Mme Queen of Heaven again told Mme Ets. Dupont how anxious she was about their cargo being held up; they were bound for Chalon-sur-Saône, and the consignee was impatient. It was a harsh world, sighed the Queen of Heaven.

It was also a harsh world on *Torne*.

'Look,' said Bill, as the lad went forward cuddling a brown bundle with long ears, 'they've got a pet rabbit.'

'I'm not too sure it's a pet,' said Laurel doubtfully. Sure enough, the skin was stretched out on the deck to cure a short while later, and the smell of rabbit stew and onions wafted across to *Hosanna* that night.

Laurel made bread later that afternoon, as she had missed the van that morning, and it seemed to be the only thing the Ets. Dupont did not stock. On a long sea voyage we are quite used to the shore-bought bread running out on the tenth day, and baking our own for the rest of the cruise. It is a cathartic and enjoyable thing to make bread. The transformation of an unpromising lump into a golden loaf is magical. Laurel's bread-song goes:

> Oh, lump of dough, I hate you so,
> You're Sickness, Poverty, Mistrust;
> You're Crack, you're Crime, you're Waste of Time,
> You're Bigotry, and Bile, and Rust.
>
> I turn you round, I knead and pound,
> I roll and fold, I form and mould.

A change occurs, a sticky mess
Becomes a perfumed silkiness.

Oh risen dough, I love you so,
You're Springtime, Tolerance and Bed,
You're Gold, you're Good, you're perfect food,
A crusty loaf of country bread.

Another impromptu meeting of the shop stewards
gathered on the quay at dusk, and Bill joined it as usual.
The canal engineer, M. Bernard, brought news that they
were ahead of schedule, and that if all went well they hoped
to start the navigation for southbound traffic the next day.
The northbound – the two barges that had been loading at
the silo quay – would have to wait a little longer. No
promises, but those concerned should be ready.

We dismantled the carpenter's shop, and put the paint
and varnish away. Throughout the flotilla there was an air
of relief. When the Queen of Heaven took to the airwaves
for her nightly broadcast she sounded almost cheerful. We
gathered that the barges waiting below the next lock south
would cross us, though there were not very many of these
since the word had been better promulgated on that side,
and most northbound traffic had taken a different route,
going via Nancy and the Canal de L'Est.

There was an air of restlessness the next morning, again a
misty and chilly one. The dawn meeting with M. Bernard
revealed that the work had dragged a bit; getting the lock
gates to 'kiss' so that they were watertight took longer than
they had thought. The complaints about lack of, or too
short, notice were to be the subject of an enquiry.

At nine o'clock, Mme *Torne* was still in her dressing-
gown, and from her husband's hairy belly depended only a
minuscule garment which appeared to be a finger stall
attached to a G-string. They had learned not to get too
excited by the words of engineers.

At about ten o'clock the Queen of Heaven came on the air

and told everyone that the work was complete, but there was not enough water further down. There was some animated conversation on the radio following this, and then at ten-thirty the canal engineer drove up to the silo quay where he announced that the *biefs* were filling very slowly because the alimentary stream, the Vingeanne, was a small one, and the lake that supplied it was very low after the heat of summer.

Already the first two barges had passed the next lock, although they were now stuck in the mud. 'Have courage, the water is coming, it will not be long.'

As the people on *Torne* wished to stop in the lock to re-embark their motor car, they suggested that we go first, a friendly gesture, for the *commerce* are ever jealous of their position in the queue. We reached the lock at about twelve-fifteen, and started the long downward chain of locks that are normally automatic, but which were specially manned for this occasion to speed us all on our way.

Did we say 'speed'? We have said that the barges ahead of us were deep-laden, and the water was lower than usual. Several times the whole procession came to a halt, the Queen of Heaven would call on her namesake in the skies above for assistance, and a barge would slowly unstick itself, letting us all move on again.

There was a momentary indication of the tension in the air when a quiet French voice observed during a radio pause that there was one among us who talked too much, followed by several of the little press-button click-clicks which the bargees use as a wordless signal of acknowledgement.

The high Plateau de Langres is very sparsely populated country. Even the optimistic canal guide books show no shops or facilities in the hamlets through which one passes. There are green fields with cattle, pretty little farmsteads built in stone, and hardly any sign of life.

At five o'clock one of the locks broke down: guess who

broadcast the bad news to the fleet. A *responsable* was sent for. The fleet waited. There seemed little point in *Hosanna* hanging about like this, pootling along at half a knot when possible, or zero when not. Even when the malfunctioning lock reopened, the queue moved on so slowly that we decided to stop at the first potential mooring, and let them all get well enough ahead to allow us to resume our normal relaxed mode of travel.

We hauled into the bank miles from anywhere just before the lock called Montrepelle, lit our fire, and had a good plate of *boeuf en daube* before going to bed.

The next day dawned fine and bright, and we determined on a gentle day's voyaging. We had a good breakfast while *Torne* passed us, crew waving, and then we waited while a barge moving upstream locked through, leaving the lock ready to receive us.

The locks were still being manually worked, and one has to join in the operation. It isn't strictly obligatory – in fact it may even be illegal – but not to do so is considered very bad manners, and prolongs the time taken to lock through, which is to everyone's disadvantage. After about seven weeks in the canals, we were becoming tolerably competent.

As we came in through the open gates Laurel at the front would make ready to lassoo a bollard and tend the rope. As soon as our stern had passed the gates the lock-keeper would start to close the gate on his side, a slow process. To have waited while he did so, and then for him to walk the length of the lock, cross over and return to close the other gate would take a long time. So Bill would jump ashore, seize the winding-handle, and start to close the other gate, and jolly good exercise it is too for stomach muscles that are not as young as they used to be.

Both he and the lock-keeper would then walk to the second pair of gates. The barge would be held steady in the lock by the one rope, with the engine idling ahead and the rudder right over, so pinning her in place – another illegality (one is supposed to moor head and stern) which we had been taught by the Dutch, who know a thing or two about barges. At the bows end of the lock each would open the *vannes* (paddles in English), which control small openings to let water in or out to equalise the levels. As the *vannes* opened, the water level in the lock started to fall, and with it our home, and Laurel adjusted her rope to keep *Hosanna* in just the right place.

When the water levels were equalised, and the turmoil in the lock had eased, the two downside gates were opened, Bill shinned down the rusty ladder in the lock wall, the rope was cast off and we gently eased our way out. For upward locking the process is reversed.

As we went on steadily, along the rural valley of the Vingeanne, the canal was sometimes lined with hornbeams and pines, as well as the more usual poplars. Where the banks were bare of trees we could see rolling slopes and villages, and sometimes a faraway *château*. It was a land of hawk and heron; the huge hawks hovered and circled above the canal, plummeting down for fish. The herons were more numerous than we had seen since the Somme, perched in a tree, the branches bending under the weight of half a dozen heavy birds, or standing in the fields in tens and twelves, leaning forward in their sombre city suits, with their hands behind their backs, rather like stock-jobbers at a meeting.

The trees were taking on a winter shape and some had very few leaves left; the aspens' balled clumps of mistletoe were their only touch of green. We caught up with *Torne* again, so the benefits of our early stop and leisurely breakfast were already lost. We were going past a woody stretch and the trees were hung with bunches of mistletoe. Usually they are too high up to reach, but we had a ladder; we stopped, put our nose into the bank and Bill climbed up to gather some of the pagan weed, as it did not seem long to

Christmas, and we let our slow companion get a bit ahead.

We soon caught him up again. We had a short consultation with the keeper of St-Maurice lock; there were no other boats behind us, and we would inconvenience no one if we stayed in the lock while Bill walked as far as the nearby village, which had a baker. He did so.

It was about a kilometre to walk along a pleasant country lane. The village seemed to be totally asleep, and there was no sign of a bakery. The small bar was deserted, and there seemed to be no one to ask. He walked over a picturesque stone bridge with some beautiful trees overhanging the small river, but there was no sign of life. Even the dogs were too tired to bark.

He returned towards the bar and was nearly there when the postman drew up.

'The baker? But the bakery is behind the bar, Monsieur. See, there where the dog sleeps.'

Of course.

Thus provided with a *pain de campagne*, country bread which is not so fine and delicate as the urban *baguette*, but keeps better, he returned to *Hosanna*, still in the lock, and we got under way once more, only to catch up *Torne* again. We stopped at Villeneuve lock to buy some of their fresh goat's cheese, have a little lunch and waste a little time.

The lock cottages were slightly different here, and presented their long side to the canal. This gave room for a window on each side of the front door. On the steps of this door at Pouilly the *éclusière* was sunning herself in a bright yellow shirt, and glorying in the number of barges that had gone through that day. She had not seen so many for a long time. Her charmingly mannered Basset hound made a pleasant change from Alsatians. He walked with his front paws at ten to two, and was called Flambeau de l'Ecluse, the Flame of the Lock.

It was a sunny day for late October, but the hardiness of

the canal folk on cold days had often amazed us. The *éclusières* were still wearing summer dresses and an apron, occasionally a cardigan on a chilly morning. Their legs were often bare. But, however hardy, theirs is not an easy life when winter comes; one lock-keeper told us that the ice needed breaking five or six times a day to keep the gates free, and that she had suffered two hernias as a result.

But now we had dropped back a fair way and by going gently we had an undisturbed run down to Licey. At each lock we had been greeted with the question: 'Are you the last?' or 'How many more behind you?'

'None,' we said. 'We are the last,' and the lock-keepers could go and relax.

At Licey there is a grassy quay next to the village bridge, well away from the lock. It was occupied by some sheep with bald shoulders, which they had acquired by wriggling under the gate of their compound; we watched them do it. Of course there were no longer any bollards, but there were some convenient trees. The only way to the road was through a timber yard, and was very muddy. But the guide books said 'restaurant', and it was time we had a night out.

We said 'Good evening' to the man who was testily herding his sheep back into their field, and proceeded through the timber yard.

Guarding the short road into the village was a dog that made up in belligerence what it lacked in size. French houses, like French lock cottages, frequently have guard dogs of a ferocious mien. One is often barked along a suburban street by a relay of dogs throwing themselves at the fence as one approaches, slavering at the muzzle in their eagerness to eat English flesh, howling with disappointment as they pass you on to the next dog, already working itself up into a frenzy. And so on down the road.

Every now and then, instead of an Alsatian, comes a fancy poodle (who makes the same sort of noises) or a dog so diminutive that one wonders if a Japanese has planted it, pruned and tended it for years before selling it to accompany a miniature bonzai tree. To make up for their stature, their barks are treble and in triple time. Given a variegated hedge of dogs, the effect is similar to several male-voice choirs singing with a boy soprano. The basses are singing 'Take me out to the Ball-Game', the baritones 'Trees', the tenors are doing fine with 'I Love a Lassie', but the boy soprano is singing something by Gilbert and Sullivan, and all without a conductor.

This little dog of Licey was barking solo apart from a tied-up animal some distance away, booming at intervals off-stage. *Le petit* was as mobile as it was small. It went straight for the ankles, which was the only part it could reach, even on Laurel, who is barely five feet tall. As we were leaving the timber yard Bill had providentially picked up a baulk of wood, and as the dog connected he lofted it in the direction of square leg. By the time the dog hit the ground it was singing a different tune, and ran home, baying falsetto. Its owner appeared in a doorway, and a pair of baleful eyes followed our progress down the road. Not a good start to a *bonne promenade*.

The village is old, and small. Very agricultural. Barely a soul to be seen. Near the canal was the village square, in reality a triangle. To one side was the Restaurant-Bar de la Marine, whose charming courtyard, shaded with creeper, had been pensioned off for the winter, the stacks of chairs roped together under covers.

Unless you were Madame *Hosanna*, you had to duck under a low doorway to enter the bar, where everything was neat and tidy. There were a couple of local people drinking, a few posters advertising the sort of events that one finds in any village and its surrounding neighbourhood in winter: 'Quine' (Tombola), whist drives, football and such, together with fearsome notices about the suppression of drunkenness among the young.

Certainly we could have a meal, but, one excuses oneself, at this season it would be *table d'hôte*.

'That's fine with us,' we said.

'At the seven then?'

'*D'accord*, at the seven.'

We returned at the seven. Behind the doors a baby was crying, and several family came and went.

'Are you from these parts?' we asked the owner.

'No, from Marseille,' he said. One of the customers remarked that this was evident; everyone knew the Marseillaises were barely French at all, and truly we spoke French better than the *patron*.

With further bantering good humour we had a couple of drinks and waited for the cry of '*A table!*'. When it came we were ushered not into a restaurant, but into the family living-room. There by a roaring stove sat Grand-mère, and several others of the family watching television. In the centre of the floor was a playpen containing a one-year-old, who had stopped crying and was now in a state of poly-activity, and in one corner of the room was the dining-table, a genuine *table d'hôte*, neatly set for two. The shaggy mat on the floor rose, and wagged its tail.

As we ate our *hors d'oeuvres*, Madame, beautiful and black of eye and hair (she was from Corsica) occupied herself on the other side of the doorway, conversing the while with the family, for the film was an American soap opera that could easily be absorbed at the same time as cooking and conversation. Our main course was served, a veal *papillote*, a better dish than we had expected in the circumstances, and Madame sat down next to Grand-mère. Both knitted rapidly, and exchanged gurgles with the baby.

We found ourselves speaking in low tones, if at all, as the atmosphere didn't encourage lively conversation. Fortunately the baby provided some sort of cabaret. We lacked a little attention when we had finished the main course, because he had fallen on a hard toy and had to be

comforted. Then the soap opera was over, a serious programme started, and everyone searched for other diversions.
Madame brought us our ice-cream dessert, and there was a
general conversation about the village.

Once upon a time the passing *mariniers* had stopped and
given their custom to the bar, but no more. There was some
activity in the summer from the little cruisers which tourists
could hire, and which would stop here for the night. Life
was calm these days.

On the wall there was a picture of the village square with
a magnificent tree in its centre. Unfortunately, the tree, a great
elm, had died of *la maladie*. It had been a royal tree, planted
personally by the King of France at some long-distant time.
The restaurant had once been called Le Vieil Orme after the
tree. Everyone had been desolated by its death. Poor Licey.

We took our leave, and began to walk back. Bill had left
the baulk of wood outside the bar, and picked it up as we
stepped out into the dark. The little dog was waiting for us,
growling and grumbling. Bill showed it the wood. It kept its
distance, but followed us to the woodyard gate, clearly
prepared to give its life should we do whatever it was that it
did not want us to do. Bill, who has sometimes a twisted
sense of humour, shook the baulk at the dog as we regained
the gate, sparking off a volume of barking hysteria that
switched on every other dog within five kilometres, and
had irate heads popping out of doors in every cottage,
whose owners added to the din by shouting at the noisy
dogs. There followed about two minutes of complete
cacophony before the village went back either to sleep or to
the television, and the goddamned English went back to
their boat. The rest of the night was silent, but for the gentle
sound of tearing grass. The sheep had wriggled under the
gate and on to the quay again.

At Oisilly the following day a touch of the South came to
meet us on this cold plateau: Laurel saw her first oleander,
or laurier rose, beloved of Mediterranean gardens,

marking our progress towards the sun. We were rapidly descending the southern slopes of the Plateau de Langres, and felt we were making progress, which we had to do because family were talking of paying a visit and it would be necessary to find an airfield for them to land on.

We were running out of provisions. The establishment of the family Dupont in Villegusien, though stocked with a variety of goods, had been unprepared for nine extra families, and two more days had passed since then. At sea, where there are also no village shops, we carried large stocks of non-perishable edibles, but we did not expect to have to provision for a trans-Atlantic voyage before navigating the canals in central France. The guide showed a shop at a village called Renève, and once again a friendly lock-keeper allowed us to stay in the lock while an expedition was made. This time, though, he could not be sure that no other barges would come in the opposite direction, so it was necessary that one of us should remain to clear the lock if on-coming traffic should appear. Laurel debated whether the prospect of a half kilometre each way was worse than the prospect of coping with *Hosanna* in an awkward bit of canal, for she was not yet confident about handling the barge alone, and decided on the walk, setting off with her shopping bag and walking-stick.

As with the bakery at St-Maurice, the shop was hard to find, and when at last she ran it down the front door was closed. There was, however, a side door, and this she entered, to find the family busy chatting. To her astonishment and pique, they refused to serve her as it was one minute past twelve; they told her she must return another day, because the shop would be closed that Saturday afternoon.

Laurel limped out and slowly returned, cross, tired and foodless.

At the next lock we passed a very friendly Dutchman, in the barge *Piu Allegro*, home port Rotterdam, on his way from Marseille to Leuwarden in Friesland, a very long run, and a good contract for a homeward-bound boat. No wonder he was cheerful.

Then after passing another *bureau de contrôle* (and having our Union Jack moved on once more by the *fonctionnaire* in Paris) we came to the traffic lights at Heuilly junction, the end of the Canal de la Marne à la Saône and the point at which it joins the river Saône. The lights were red, and they stayed red for a full eight minutes while a big barge moved clear, then we were given the green light into the wide river Saône, saying *adieu* to the little canals of northern France. From now on, we could expect the journey to become even less gentle than the last stretch had been.

7

The Saône and so on

'E, de canau en canau, pèr la Sono, es descendu de soun pais de Flandro, Coume davalon dou neblun li céune i clar dou Vacarés, quand vèn l'autouno.'

. . . .

'From canal to canal, by way of the Saône, he came down from his land of Flanders, as the swans come down from the misty North to the pools of Vacarés, when autumn comes.'

Coming out of the lock at Heuilly and into the full broad sweep of the Saône gave us the sort of feeling that a butterfly must have on leaving the chrysalis. Instead of going slowly along a channel so narrow that one could spit on to either bank (assuming that one were ill bred enough to want to do so), the boat was now able to spread her wings as we pushed the throttles up to full ahead in a river that is more than a hundred metres broad.

There was no longer the resistance to motion caused by a narrow and shallow channel; now the motor felt free, *Hosanna* was more alive to the touch, and there was enough speed to generate a head wind.

So fine was the sense of freedom that we had barely time to note that the mooring and restaurant that we had been led to believe by the canal guide would be a good place to stop was not only firmly closed and shuttered, but for sale.

Of course the freedom was soon circumscribed. One expects to keep to the right in these channels like any other waterway; it is one of the very few conventions that is completely worldwide, and was in fact the first ever to be so agreed, the implications being that if seamen ruled the world we would all manage to get on with each other instead of squabbling. Seamen, however, are too busy being seamen, and have more sense than to meddle in politics. Pity.

On big rivers, where the floodwaters can be very fast indeed, there is also a convention that a boat going upstream is allowed to hug the inside of a bend. This is not only so that the boat going upstream has a shorter distance to travel, but, more importantly, because the current is always a lot stronger on the outside of a bend, where the flow, pushed there by centrifugal force, can be so strong that a barge cannot overcome it.

The French, being a formal and logical nation, have designated the bends where this contrary rule is to apply, and we came upon the first of these almost as soon as we entered the Saône. Anyone can foresee that there is a potential danger spot where the two conventions meet. One boat is happily proceeding down the right-hand side of the channel with the on-coming traffic passing on its left, and then suddenly the position is to be reversed: it must go over to the left side, and the on-coming traffic likewise, so that they are now passing on the right.

It does not need much imagination to picture the traffic problems if such a requirement were to be introduced half way over London Bridge, for example. It is one of those occasions when the potential for disaster is so enormous that everyone takes elaborate care, the only exception being the occasional driver of a 600-h.p. motor-boat on his way to the Med as fast as he can go, and in autumn there were not many of these.

The Saône (pronounced 'sown' as near as an Anglo-Saxon can manage) is a canalised river that has recently been upgraded to take the big barges of 1250 tonnes. Its lock system has been changed; now there are fewer locks, each with a bigger drop, and they are enormous compared with those in the Marne canals. The locks are all operated by power, and controlled by men high up in towers resembling the control towers at airports. They look down on the passing traffic, and if necessary address it via loudspeakers that are largely unintelligible, just like those at airports.

There is no more social interchange between *marinier* and *éclusier*. No more eggs and vegetables for sale, no goats or rabbits in hutches, no hens scratching by the lockside. The pots and tubs and flowerbeds are replaced by gravel, and sometimes a single architect-designed tree, which has succumbed because it was someone else's job to water it.

The environment is quasi-military, impersonal, but not noticeably more efficient.

The locks now keep watch on VHF radio, so that it is possible to call ahead and forewarn them of one's arrival, thus lessening the chance of having to hang about while the lock is prepared. Such is the cost of each operation that pleasure craft are normally expected to wait either until they can lock through with a commercial craft (for the locks are big enough to take several barges at a time), or until a small convoy of pleasure craft has gathered. The rules say that pleasure craft can pass singly if the wait exceeds half an hour, and as in winter there are not many pleasure craft, most lock-keepers let one through without unnecessary delay. It seems that if forewarned by radio the waiting time starts earlier. But it is because of this official waiting period that, from the Saône onwards, pleasure craft begin to travel in convoy and thus get to know each other. At the moment we were still alone.

Rebuilding the locks, an enormous project, has also changed the levels of the river, and this has affected the possibilities of mooring. In several localities what was once a good quay is now either high, dry and unapproachable because the water level has been lowered, or has disappeared under water because the level has been raised. In most places the level is lower, and a stratum of tree roots, fishing staithes and steps down to landing stages, all now on dry land, indicate where the level used to be.

The river Saône rises up in the Vosges mountains and, coming down from Lorraine, meanders through the plain which is the only natural north–south route in Europe, the valley between the Jura mountains and the Massif Central. This route therefore carries several motorways and railways, as well as the Saône. From Bronze Age times it has been the amber road from the Baltic, the tin road from Cornwall via the Seine, and the salt road from Italy.

Now we followed the river, wide and winding, through this windy level plain, dotted with Friesian and Charollais cattle and patched with fields of tobacco, potatoes and sugar beet. This low-lying country is protected from flood by *levées*. Talking to a river engineer, we were surprised to learn that a river tends to raise its own *levées* as it winds across a flat country, but these are not high enough to cope with floods and in the valley of the Saône winter and spring floods are a big problem. Many of the riverside houses are built up on stilts or banks of earth, and one wonders about the prudence of those who have built directly on the normal ground level. The chances are that it will be these people who will write all those letters of complaint to the papers that follow every flood, every winter.

At intervals the engineers have by-passed a particularly sharp bend, and this usually means that a town or village which was at one time on the banks of the Saône is now on the banks of an unnavigable and shoaling backwater.

The first town with an approachable quay is the old city of Auxonne, where, under huge plane trees, its inhabitants built their quay in the form of a wide flight of steps, which made it a little difficult to snuggle against but quite easy to get ashore, once one had uncovered the steps from under the immense drift of autumn leaves which had turned them into a shifting golden ramp.

Joy of joys, Auxonne is noted for an excellent restaurant,

so we put on our glad rags, entered the town through the immense gate in the old walls and, after a provisioning expedition, went in search of the good and gluttonous life at the Hôtel de Corbeau.

The cuisine was not at all bad, but neither was it outstanding. The wine list was, however, disappointing. There were some excellent wines on it, but as we looked at the cheaper end of it, where we belong nowadays, we found to our dismay that ordinary and commonplace wines were offered at prices that can only be described as exorbitant.

We have said that the price of a meal is extremely good value in almost every French restaurant, but now and then one comes across examples of profiteering in the wine list in a land where wine is not dear because of any government rake-off, but only because of man's greed. This applies both to the *vin de table* at the lower end of the scale, and with even more force to the *grand cru classé*. At this restaurant, for example, a top-class claret was priced above that for the same wine of the same year at the Savoy in London, where both taxes and overheads are considerably higher.

But we had a good dinner, as dinners go, and we were ready next day to travel the short distance to St-Jean-de-Losne, which is the busiest barge town in France, with the exception of Conflans-Ste-Honorine, near Paris, which latter is the constituency of the French prime minister, M. Rocard, not that that helps the hard-pressed *marinier*, as several of them said.

The town of St-Jean is small, with only some 2000 inhabitants, but it is at the junction of the river Saône with both the Canal du Rhône au Rhin, which is very important, and with the Canal de Bourgogne, which is less so. Here the routes from England, the Low Countries and Germany meet the route to the Mediterranean.

The river is very wide, and crossed by a single road bridge joining the two halves of the town, much of which is spread along the banks. At river level one's first impression is of moored barges, dozens of them, for here is one of the major *bourses* where work is offered. The *bourse* is where the commerce goes to look for its freight. They are also called *bureaux d'affrètement*; *fret* being the best the French can do with the word 'freight', which contains impossible consonants for them. The *bourses* are linked to each other by computer these days, but there is still a big chalkboard on the wall with the week's voyages marked on it, so on entering one can see what is going on.

A barge which has just completed a voyage and discharged her cargo will at once have her name put down at the nearest *bourse*. When she comes to the top of the list, and this may take weeks, as we have seen, she is offered the next voyage that comes up. If the cargo, distance or destination is not to the captain's liking, the voyage will be offered to the next in line, but the first barge does not lose her place at the head of the list.

Only a French barge may take cargoes between two places within France. Belgian and Dutch barges (and Germans, but these are more rarely seen in these waters) must, therefore, wait for a cargo that will take them out of France; a typical voyage would be Lyon to Bruges, which might take eighteen to twenty days, fully laden.

Unpopular cargoes are those which are dirty and cause a good deal of cleaning after unloading, or those which 'eat the boat' and cause it to rust, like sand, gravel and potash.

On the upstream side of the bridge there are berths for large passenger vessels which still ply, though no longer on scheduled services; now they run trips for tourists. Although the day was fine for October most of these had gone into winter hibernation.

On either side of the bridge on the St-Jean side there are shipyards and fuelling stations, and downstream of the big bridge, almost hidden alongside the entrance to the Canal de Bourgogne, is a narrow cut under a bridge leading to a big basin, once the port of St-Jean, but now the home of several near-derelict barges and the *port de plaisance*, where the pleasure craft are expected to go.

There were about fifty or so small craft, mostly forlornly abandoned for the winter by owners who lived elsewhere. In addition we came across a considerable number of barges which had been converted for habitation. Some were 'hotel' craft offering holidays cruising the canals and rivers in comfort with a professional crew to do all the hard work, both navigationally and domestically. Others were floating homes like *Hosanna*, and in this case it is necessary to distinguish between those that had become permanent house-boats, their owners abandoning all thoughts of travel, and those like ours who had owners suffering from wanderlust.

The English word house-boat has entered the French language, undergoing both a change of spelling to *hous-bot*, and also a subtle change of meaning; to them it seems to mean any boat on which one may sleep.

We had the impression that there were not many like us at St-Jean, though there were several boats of varying quality for sale for the purpose.

In general, most of the boats converted for a cruising life are of Dutch origin, as *Hosanna* is. There are several reasons for this. One is that the standard French *péniche* is thirty-eight metres long, and apart from hotel barges (and several of these are converted from *péniches*), this is far more than one needs for a home, and is troublesome to maintain and handle. Another factor is that the French have tended to build their boats for a comparatively short working life, and many of them have thinner steel plating than the typical Dutchman. Canals in France are less well maintained than in Holland: a barge spends more time ploughing through mud and sand which abrade the plating, especially on the starboard side, for barges keep to the right when passing each other and their right-hand sides are often scraping along the sand and rocks that line the canal. It is always the right-hand side of the bottom that needs renewal first in any barge.

In Holland, on the other hand, there has always been a tradition of building up to a higher standard than the minimum, and a pride of craftsmanship that dictates that a vessel, no matter that it is 'only a barge', must look good too. As a consequence there now exists a large number of comparatively small Dutch barges which serviced many of the less industrial parts of Holland in the days before motor transport. Some of them are more than a hundred years old, and still in excellent condition. One can always tell them: like *Hosanna*, they have more pointed bows, and long, sloping, elegant sterns, called counter sterns.

The Dutch government became anxious a few years ago to upgrade the size of the average barge in Holland to improve the efficiency of their waterways, and offered to any owner of an undersized barge a subsidy to build one of the new large ones, provided he sold the old small one out of the country, or scrapped it. The result was a buyer's market in small Dutch barges, for the subsidy offer was a generous one, and Dutch owners could afford to sell their tiddlers for almost a song. We ourselves had bought *Hosanna* (then sailing under the name of *De Tijd Zal t'Leren IV*, which we could not pronounce convincingly) for little more than the value of her engine, and she had been in such good condition as to astonish the British surveyor who was engaged to check her over. We had sailed her across the North Sea to Great Yarmouth in the middle of winter, and had acquired confidence and a burgeoning love for our vessel.

When one marries a boat it is just as serious an affair as marrying a person. It is certainly true that some people buy boats and sell them again like motor cars, but when one falls in love with a boat it is not easy to divorce. When we sold *Fare Well* after more than ten years of living in her it was like being bereaved. Two years later the pangs are still with us, and when the new owners wrote that they had suffered damage in a bad storm we almost wanted to fly off and be at her dockside.

Sentimental nonsense? Depends on your point of view; one lives far more intimately with a boat than with a house. Of course a boat is not a living thing, nor should one wallow in anthropomorphism, but when living in a boat one becomes aware of the purpose and strength of every little piece of material which has been melded into a coherent whole to provide safety and shelter against conditions no other form of residence has to endure. One respects both the material, the components and the men who put them all together.

Every small-craft seaman will be able to tell of a time when he was saved from death not by his own skill, but because his boat performed better in survival conditions that he could reasonably expect. A mast perhaps, or some other part, underwent a strain that should in theory have destroyed it, but it held, by the grace of God (for we all become believers at times like that, even if it is only temporary). As a result of such experiences one forms a relationship that is not bound by common sense; a kind of love is felt for the inanimate object that saved one's life when it didn't have to.

It is possible to see how, before the days of the one God Invisible, people could make inanimate or even abstract things into gods, when such objects were so much part of their existence, and when they so often depended on the apparent goodwill of such an object. The sea is the last place left on earth where man is so absolutely dependent on his home-ship that the relationship transcends scientific reason.

But we have digressed. Let us go to see the agent for the sale of these barges, a multi-lingual German called Gerard running a business called H_2O; this was in that state of amicable disorder which indicates that one is dealing with nice people. He also has a very interesting secondhand nautical goods shop, a rare thing in France.

There was no mooring for us in the basin. It is a feature of *ports de plaisance* that berths are limited to very small craft only, and that even then they are semi-permanently occupied by boats that are used once a year, making them into a nautical long-term car-park. This was not exactly the case at St-Jean, but all the reasonable possibilities open to us were occupied by barges for sale or laid up for the winter.

We moved out of the basin, and to our relief were welcomed alongside the commercial barge *Energia*, from Béthune, where we were soon enjoying a welcome cup of coffee in the immaculate day cabin of Jeannot and Georgette Timmerman, with its enormous oil-fired stove that heated the whole accommodation area, and which it was impossible to approach without being grilled alive.

We began to realise why barge ladies were still wearing thin summer blouses, as the cabin was toasty cosy and we had to remove our winter jerseys. Their day cabin, with flowered paper on the walls, contained a sofa, a carved sideboard holding photographs on a crochet runner, and a cabinet of ornamental china. The table round which we sat was sensibly covered with oilcloth, as the children were playing a game. The two bedrooms opened off the day cabin, as did the galley. *Energia*'s galley was panelled in wood, and had a stainless-steel sink, a fridge and a gas stove.

There was a hi-fi, and a TV set. One of the features of any commercial barge's exterior is the huge TV aerial it carries – a thing so complex, so full of wires and parabolas, of forks, rakes and metal prongs pointing in all directions, that you feel it might beam down the moon if you pressed the wrong button. Many of these antennae are mechanised, and can be seen eerily turning and elevating, seeking the best reception for the evening.

We admired their beautifully neat and compact wheel-house, with Madonna and St Christopher medals built into the dashboard, a loudspeaker to the front of the barge (thirty-eight metres is a long way to shout in a high wind), and the jungle of greenery and froth of lace curtains without which no barge is complete.

We knew by now that washing facilities, for both clothes and people, were almost invariably right up in the bows, in the *salle d'eau*, or water room. We vividly remembered our first, spartan journey in an unconverted *Hosanna*, in the depths of a Dutch winter, when we had discovered that there was no *salle d'eau* at all, not even a loo. Laurel had brought a lavatory brush in our journey pack, and we had nothing to use it on. We bought a bucket at the next village we stopped at, and that was how things remained for a long time. The bathroom was not the first thing that was installed on conversion, as other more serious works came first, and we had to use the boatyard facilities, an outside gents, for some time. While this was extremely clean, it had been disconcerting to find it sometimes occupied by a large sheep, who grazed on the marsh outside and would come in for a drink from the toilet bowl.

It can be imagined what a relief it was when we finally got our bathroom working on *Hosanna*, no matter that the bath is still not panelled in, wires droop from the ceiling and the cupboards that will one day house the contents of the cardboard boxes marked 'Second Aid', 'Cat', 'Things for colds' and 'Useful in bathroom' are not yet built.

The crew of *Energia*, though they were French nationals, sailed under the Belgian flag; they felt it gave them a better chance of finding voyages, and seemed to prefer the long ones up to Bruges and Rotterdam. They visited the *bourse* each day in hope of a voyage, though few were coming at present, and many barges were waiting.

The children did their lessons by post, and were a month behind because of the postal strike. Once all the barge towns had schools for the barge children, but a notice in the Bar de la Marine regretted that if the numbers of scholars presenting themselves dropped below fifteen the school would have to close, and the children would then have to go to school at Chalon, seventy kilometres away.

The post office still had no mail for us. The engineer at Villegusien had kept his promise and had sent a car to Rolampont for our mail, but there had been none there either, and he had arranged for it to be sent on to St-Jean for us when it did arrive. There should therefore have been two batches at St-Jean, and being somewhat troubled we tried to phone our daughter back in England, only to find the phone box out of order.

'That is normal,' said Jeannot of *Energia*, 'you should visit the Bar de la Marine, for there you will find a telephone which accepts coins.'

Going ashore from *Energia* entailed a vertiginous walk across their long springy gangplank over a chasm with cold-looking water at the bottom, a bit unnerving for the first few times. We also had to scramble up to *Energia*'s deck in the first place, since being empty she rode much higher out of the water than we did, sufficiently so to discourage the kitten from attempting to change boats.

The Bar de la Marine was not, to our surprise, one of those French bars that are occupied only by men: it seemed that the barge ladies also called in with their children; it was a sort of club. The main activity was at a table in the centre where a game of cards was being played with great vigour, the cards being banged down on the table each time as if making a vehement claim to instant victory, and the crowd of eight or ten spectators buzzed and kibitzed with enthusiasm.

The telephone was on the staircase windowsill, and therefore ideally suited to a person with one leg longer than the other. It worked well, and enabled us to start the arrangements for our big party.

Bill was coming up to his sixtieth birthday, which would almost coincide with our daughter's birthday, so, feeling in need of some family sociability, we had arranged a reunion for the second weekend in November at a place yet to be chosen a little further down the Saône.

It was already 1 November, *Toussaint*, a public holiday for All Saints when it seems as if everybody moves to the local cemetery for the day, all shops except florists are closed, and millions of chrysanthemums are sold. The precise number is published in the papers the next day, like the number of persons shot on the opening day of the hunting season.

The French are very fond of figures; the radio news bulletins are full of them. The figures should be as large as possible, preferably in millions or milliards, and to achieve this the price of many substantial objects (like boats) is given in millions of centimes. After about twenty years of the new franc one can still suffer near heart failure when some old lady presents one with a bill in old francs, mostly consisting of enough noughts to make a bead necklace.

It was a very cold night, and so the weather stayed for a few days. After attempting once more to obtain our mail, and stocking up at the local supermarket and an excellent *charcuterie*, we set off downstream, passing the special three-level quay which the barges use for loading or un-loading the cars that nearly all of them nowadays carry on deck. Some of them use a small hydraulic crane for this purpose, but the majority seem to position the car across the barge, and when they need to disembark it they find a quay of roughly the right height and push the car off on two large planks. The demand for this facility at St-Jean, where twenty-four barges were waiting up to four weeks for a cargo, was such that a special quay had been built.

We did a long run that day, sixty-three kilometres with only two locks. We were covering the ground much faster; at this rate we soon would catch up the swallows.

Chalon-sur-Saône, where the Canal du Centre joins the Saône, was crowded with barges and we could find no mooring close to town. We passed *Regina Coeli*, busily unloading her cargo, and Madame Queen of Heaven called us up on the radio, more cheerful now, to wish us well. We ended up berthing fourth one out alongside some commercial barges, but made no social contact on this occasion. We went for a walk in search of a meal out, but there was a bitterly cold wind blowing and the prospect of a difficult clamber over three barges to our precarious berth would certainly have spoiled our enjoyment of the wine. We lacked enthusiasm, and as we returned to walk back along the tree-lined avenue we heard a terrible sound – cat up a tree plus pig being tortured with elements of metallic rasping, hard to describe and very disturbing. We stood still, looking around for the source of the noise, which was coming nearer, and were aware of locals regarding us with amused smiles. They knew. Every evening at sundown a loudspeaker van paraded the avenue broadcasting a magnified alarm call to scare the starlings away. The cure seemed worse than the disease.

No curfew tolls the knell of parting day,
The peace of Angelus is here unknown,
Across the *boulevardes* of Challonais
The strident hoarseness of a megaphone.

The starling scarer comes, to end the calm,
Broadcasting fear and warning all around,
A million starlings shrieking their alarm
With much High Tech to magnify the sound.

It works! A thousand starlings screech and rise,
Dropping their offerings in the tourists' eyes.
For half an hour a-wing the starlings stay,
But aren't the tourists also scared away?

Ears ringing, we returned on board, sponged our jackets, and ate our pork chops *Avesnois*, a recipe from the North of France, simple to do and always appetising.

Before we left the next day we had to free a huge tree trunk wedged across our stern by the current, the sort of thing that *éclusiers* hate to find in their locks, but often do after flooding.

We had better luck with both quay and restaurant at Tournus. The bridge across the river had been recently re-built, and a new quay had been made for the construction materials to be unloaded. There was plenty of room to moor right there in the town centre, and the guide books told us that here was one of the best restaurants in France.

A glance at the menu outside showed that it would cost us a month's housekeeping to eat there. Fortunately, in any town where there is a gastronomic giant, a little of the expertise and enthusiasm seems to rub off on to lesser establishments. We chose a small hotel-restaurant, Les Terrasses, on which Mr Bibendum had conferred his little red 'R', which means good food at moderate prices, and here we enjoyed one of the best meals we have ever had in France. Perhaps we should take the time to describe it.

We had a 'warm' salad with chicken livers, some snails, and then *ris de veau* (sweetbreads) with *morilles*, and then a sorbet made of green apples served with a glass of Calvados to pour over it. We drank a bottle of good Aligoté. There was not a single thing we could fault.

We think that one can usually eat better having the speciality of a moderately good restaurant, than going to a highly prestigious place and having its cheapest offering, which may well have been cooked by the apprentice. In the case of Les Terrasses, what we had was by no means the top of their bill; it cost us 304 francs, or about £28, more than we usually spend, but we were starting to celebrate a birthday, which in our case is an excuse for several days of indulgence.

We landed our little car at Tournus. This had not been planned, but we were by now approaching Lyon, with its airfield, and we needed to find a stopping place for the party. The quay at Tournus was exactly the same height as our *roef*, so it occurred to us to seize the opportunity, and to prospect for a good riverside *auberge* by car.

Rolling the car off on two planks as the bargees did sounds easy; in fact it was trickier than it looked. The car gained momentum as it was slightly downhill to the quay, bounced over the baulk of wood placed there to stop it, and would have gone on driverless but for the preventer that Bill, always a belt-and-braces man, had fixed to the back bumper.

The thought was not too far away that Les Terrasses might be the best spot for some family gluttony. We phoned our daughter.

Over a lifetime subtle changes take place in one's relationship with daughters. For the first few years they are completely dependent on their parents, and for a few more they accept parental authority. Then they start to question that authority, until as they become independent they toss it aside altogether. There is perhaps a brief point where one loses a daughter completely.

Gradually, if you are lucky, the relationship re-establishes itself, and becomes a friendship between equals, each recognising that the other has certain areas of expertise. Then subtly one finds that one's daughter is giving one advice in the sort of way that means she expects it to be followed. From there it is but a short step to being told what to do, and the process ends up with daughter making decisions on one's behalf, hiring the ultimate nurse, and shipping you down to The Cedars, The Poplars or The Elms. However, we are not there yet, and are enjoying the pleasant stretch in mid-life when we have stopped worrying about our children, and they have not started worrying about us. Except – a twinge of firmness is creeping into the advice. We'll have to watch that.

Bill felt that this time he was being suggested to rather more firmly than he was used to. Is this what being sixty is all about, he thought – a watershed across which there is no return? He came back from the telephone box having accepted Shelley's suggestion to move nearer to Lyon, and wondering why he hadn't argued. She had told him that one of the Sunday supplements had been very enthusiastic about a small hotel by the Saône at a place called Crêche, and she liked the sound of it.

Dutifully we drove off to check this out, and eventually found a riverside *auberge* with moorings. We reported back to the Commanding Officer in England, who accepted the results of our reconnaissance with good grace, and, with permission to proceed, *Hosanna* headed for the rendezvous at Port d'Arciat, where the river was closest to Crêche.

On the way we stopped for some shopping at Mâcon, where one can moor alongside the market place, and a jolly good market it is on a Saturday morning: a huge area by the river under large trees, where colourful stalls of all descriptions echo with a great deal of noise and bustle.

We bought some treats as it was Bill's birthday (the 'official' birthday would follow the next weekend) – salmon steaks at the fish stall, some goat cheese, fresh fruit and vegetables. Many of the vegetable stalls are Spanish, and truck-loads of fresh vegetables come from Spain every day.

The guide books counsel against mooring in the town centre of Mâcon (there is a *port de plaisance* about two kilometres outside town), saying that the town quay is *très mouvementé la nuit* (very lively at night). This warning fascinated us and we had to find out the cause. Was it the noise? Pneumatic drills? Lorries turning? Barges loading? The rehearsal rooms of the Insomniacs Brass band?

There was no sign of anything untoward on Saturday morning. A few cabbage leaves floated down from the market, but nothing that you would not expect. We consulted the *bateliers*, who were all moored up at one end of the long quay; the WC was mentioned, rather coyly, but no further explanation was forthcoming.

It was *Regina Coeli* who provided the answer – the King in his woolly hat rather than the Queen, for it seemed that the lady wished to avoid the subject. Bill was taken discreetly aside. The information he was given, relayed to Laurel, gave rise to the following Swiftian rhyme:

> Renowned MÂCON, thou whose vines
> Do furnisshe us with heady Wines,
> A usefull BUILDING didst conceive,
> The calls of Nature to relieve,
> And placed it, private and alone,
> Close to the MARKET, by the Saône.
>
> Alas, not onlye flocked thereto
> Those who had quaffed the Premier Cru,
> But from the Marches of all FRANCE
> There tittupped Regiments of *Tantes* –
> Converged on MÂCON, where the quai
> At night is hight *très mouvementé*.

Be wary, Childe, be goode, and hide
From fairies by the riverside.

The nightly comings and goings can be disturbing to
boats as the moorings are in close proximity to the convenience. The police keep very well away. So one end of
Mâcon quay is Cottage quay, and the *bateliers* keep to the
other end if they can.

After lunch we steamed on to the party rendezvous we
had chosen at Port d'Arciat.

All along the banks of the river, and also along the canals,
are little spots marked on the maps as *ports*. Mostly these
have reverted to nature, and there is nothing to be seen, for
they are the village quaysides where barges in the old days
would deliver the goods the village needed and take away
the produce they wished to sell. In places there remain the
vestiges of an old quay, but at Arciat, which seems to be a
virtually non-existent village, the stone quay was in good
condition, and about fifty metres long.

There was a barge moored there, an old type that was
well kept but up for sale with little hope of finding a buyer.
Being empty, it had been moored well up towards the end
of the quay, where it was shallow, and even the short,
exposed end of quay had insufficient water for us to swing
our stern in. We spent most of the afternoon berthing and
reberthing, trying different approaches. In the end we
moored with our anchor out in the river and our stern
pinned in tight alongside the old and nameless barge.

It wasn't a bad mooring as moorings go; the big barges
romped past about a hundred metres away, and we could
just feel their wash.

The quiet birthday dinner *à deux* on board was mostly
good, though we failed with the warm goat-cheese salad –
the cheese we had bought at Mâcon turned out to be
rock-hard. Some people prefer it like that, it seems; we
should have explained what we wanted it for. The salmon

steaks with sauce *à l'oseille* (sorrel sauce) were delicious, and after some pâtisserie (bought, as there were definite limits to what the galley could achieve in its unfinished state) we had coffee, Bill's favourite marzipan and liqueurs, and turned in feeling content and rosy.

Noises awoke us early on Sunday morning: low voices, the squeak and thump of oars, an exclamation, a chuckle of water. We looked out into thick fog, and could dimly see close to us several fishing boats containing the undaunted French at their favourite Sunday pastime, wrapped in arctic clothing, breath steaming in the fog, catching fish they could not see.

As the unseen barges passed we could hear the roar of their engines, note their passing by plaintive wail of syren, and wait the thirty seconds for the waves to reach us.

The quay was grassy, and comparatively unsullied by dog. Later in the morning a weak sun dispersed the fog a little, and showed us a large barnlike building with faded lettering, barely readable, saying it was the Auberge de la Marine, special parties on command. A relic of the old days, where barges used to stop for the weekend and have balls and special dinners.

Beyond that lay the Relais de la Saône, an old building with a smart little restaurant, and a modern annexe containing bedrooms, grouped round a courtyard overhung with chestnut trees. At this time in mid-November there were few customers, and we had a quiet but extremely copious (we were in Burgundy, remember) Sunday lunch.

We spent a week there while the family, including ourselves, got their various acts in order. We went back to Tournus to fetch the little car that had been left behind, hitching a lift with the restaurant-owner into the nearby village of Crêche where there is a country railway station. We caught the evening train to Tournus, packed with commuters homeward bound from Lyon.

The station at Crêche was completely deserted and locked;

there was no ticket to be bought there, and no inspector or guard appeared on our journey to sell us one. At Tournus, as honest Britons, we tried without any success to find someone to pay; all the commuters left by a side gate and we were swept through with them, expecting the while to hear whistles blow, or feel the heavy hands of retribution descend on our collars. We remembered the trouble we had got into up in Picardie, when we had purchased tickets but failed to get them *composté* (date-stamped) and the inspector on the train had given us such a dressing-down that Bill had had to resort to the last refuge of the coward, and refuse to speak French.

Now we had a car to shop with, and as there were no facilities of any sort at Arciat we drove to the nearby Centre Commerciale to stock up for the weekend. Bill then immersed himself in woodwork, progressing the galley as we had visitors to feed, and making some improvements to the spare cabin.

We discussed a special menu with Monsieur Vaucher, the *patron-chef* of the Relais. He listened politely to our suggestions, and the next day came down to our boat with a draft menu. Marvellous, we told him. *Allez-y*. Go for it.

Until now our meals on the boat had been served on everyday crocks, coffee in mugs, and wine in the squat French glasses with a trade-name more suitable for condoms, which are almost unbreakable, even at sea (the glasses, we mean). Now was the time to be more elegant. Laurel hunted for the boxes containing larger bowls, saucepans and ovenware, candlesticks and damask napkins, and the butter knife. Some of these items she found, but the dozen wine goblets she knew she had in a box somewhere eluded her until months later. She also did the huge clean-up that mothers do before children visit, and children do before mothers visit, but cleaning *Hosanna* was not as easy as house cleaning, especially in its unfinished state, and caused her to turn a hair.

Dust and muddy footprints are one thing, but woodshavings, scrags of sandpaper, steel swarf, chunks of insulating foam, bent nails, rows of old tin-cans containing threadbare paintbrushes standing in jellied turps, and sticky knots of masking tape are quite another. And some of this mess was in her galley, too. But once the clean-up was done, she felt she could forget all the boring stuff and have fun cooking.

Tassie chose this moment to fall in the river while following Laurel ashore. She made a mighty leap for so small a thing, and found no purchase on the metal sides of the second barge, slithering down into the water and swimming strongly among the choked leaves and other debris. We found a rope and held a noose in front of the busy paws, which hooked on and allowed us to haul her aboard.

She was subdued for at least half an hour, and Laurel, always an opportunist, profited by her unwonted calm to douse her with flea powder. We hoped that she had learnt two important lessons that our earlier cat Nelson had learnt well: one, don't fall in; and two, if you do, find a rope. To which this kitten could now add a third: then hide, if you want to avoid flea powder.

With *Hosanna* clean and tidy, some of the cardboard boxes stowed under the floor, and some tastefully concealed under rugs, we were almost ready. We combed the paint out of our hair, washed our hands and knees well with turpentine, bathed, and set off on Friday in the rusty little old car to the airport at Lyon, parking cheekily by a grand Mercedes while we met Shelley and Nigel. It was a pinch getting four and luggage into a baby Fiat; one has to be on very good terms with one's passengers.

We have noticed in Italy, where these little cars abound, that almost all of them seem to have a wardrobe on the roof, but even without this facility one is amazed at what the Italians can get into a baby Fiat. Seven nuns, for example.

Our son Ben and his lady Claire arrived, having had a rather tiring journey driving our other car down for us.

Bill's sister had been unable to come, so with six of us the party was now complete.

The first thing we discovered was that this was not the hotel mentioned in the Sunday supplement. Of course we were committed now, but honour demanded that we search for and check 'the super little place by the Saône'.

We found it up by the railway station, four kilometres from the river, but a nice-looking spot all the same. As we contemplated it an express from Paris to Lyon roared past, followed almost immediately by a long, rattling, north-bound goods train. The subject of the wrong hotel was dropped in favour of more interesting topics.

We all dined at the Relais that night, and were introduced to a fish that was new to us, the *sandre*. This is a freshwater fish found in a few of the big rivers in France. Similar to, but subtly different from, both pike and perch, it is a great delicacy. Our *patron* explained why the French like pike and the English do not. The French catch their pike from rivers that run over sand or pebble beds, while pike are mainly found in England in the muddier rivers like the Broads, and this makes all the difference to their flavour, for their flesh absorbs the taste. French pike, and the *sandre*, have no taste of mud, the latter particularly living only in rocky streams. It is a very good fish indeed.

And then on the Saturday, we had our special dinner:

Coupe de Champagne
Escargots de Bourgogne en feuilleté
Quenelles de brochet sauce Nantua
Porcelet rôti et sa garniture
Fromage frais
Vacherin

the whole served with Morgon white wine, Fouissy-Rouge, and a Muscatel for the dessert. Our host did not stint on the ceremony (or *cinéma*, as the French term it), parading the golden sucking pig in all its decorated glory. It was a large piglet from the farm next door; the little pigs, he had explained, tasted only of milk and this would have much more flavour.

We ate and drank, and talked and laughed. Shelley had only two mouthfuls of the champagne that M. Vaucher had served on the house, and was then awash with Perrier. We were going to be grandparents, she told us. This news made a good birthday present, and Bill felt that if he had to be sixty, this was the best way to do it.

We had expected to spend the weekend at Arciat, and bid farewell to our guests there. Please read preceding pages about daughters. We found that a trip in *Hosanna* was envisaged, and that somehow without our true volition a quiet Sunday had become a shade more hectic. The Commanding Officer had arranged that we should proceed downstream to Lyon, the cars being driven down by relays of drivers.

'But the goose . . .' said Laurel (it had been bought at the famous Haven Bridge Gooseworks in Yarmouth, and had come all the way from England in our freezer). The celebrations were to continue on board, however, and she and Shelley drew the short straw for driving the Fiat forty kilometres to Trévoux, the first stop. Nigel, who is no mean cook, took responsibility for the goose, and after paying M. Vaucher an unbelievably modest amount for his efforts (the bill for the special dinner was 165 francs, or £15, per head including wines), we were under way again, though this time we had to go at full speed as our organisers had over-estimated the rate of progress. At Trévoux the car teams changed, and Laurel returned with relief to her goose and galley, leaving the only remaining lock at Couzon to be worked by the visitors.

We were slow to berth at Lyon, for the quay recommended in the guide was roped off and undergoing repairs, and we had to drift about looking for a convenient substitute.

To moor in a big city after the peace of the countryside is to learn to live once again with noise and hustle, but the hilarity of a family dinner on board drowned all exterior sounds; the goose was if anything improved by the succession of chefs who had tended it, and nobody noticed the absence of wine goblets.

At an ungodly hour on the Monday morning we managed to get Shelley and Nigel off to the airport, barely in time to catch their plane and get back to work. Ben and Claire caught the *train à grande vitesse* back to the north, leaving the car behind for our use (we now had two cars and a boat to shuffle and mix for the next stage of the journey). We looked for and found a better berth further downstream at the northern end of the Quai Tilsit, where the empty commercial barges wait alongside the *bourse* for their next cargoes.

With the departure of visitors, however much loved, a delicious langour creeps over us. We can eat what we like when we like, we can slob around in old clothes, turn the radio on for music or the weather forecast, and return to a pleasantly sluttish existence with paper napkins, marmite in the butter, and coffee in mugs again. Parties are a good thing, but they are exhausting.

For almost the first time on our journey we met our 'tramping congeners', as the Bird Park brochure had called fellow passage-makers. Moored close by were a pretty Dutch *tjalk*, *Dolfijn*, with Californians Al and Joanie aboard; an English catamaran, *Lambrusco*, housing Francis and Marion; and a small Dutch yacht, *Anya*, containing Cape Cod poet Ding and her Dutchman. We forgot that parties were exhausting and proceeded to have quite a few of them on any boat which felt strong enough to be host, the relaxed and impromptu sort of party well known to cruising people, where everyone pitches in with food and drink, track suits and carpet slippers may be worn, and nobody has heard your stories before.

We took Ding, who was travelling on a small boat for the first time, to the market. Laurel has years of seagoing experience regarding Boat Shopping, and has developed a cuisine of diabolical ingenuity with elements from the four corners of the world, combining tins, dried foods, rice and pasta with what fresh food is available to make appetising dishes. Here in the abundance of France her skills were scarcely needed, since we had time to shop and a refrigerator. On *Anya* Ding had no fridge, no time, and limited storage space, so we pointed out good food for boats carrying no refrigerator, with limited space and a meat-hungry Dutchman. We knew the feeling: cold fresh air and the work of getting a boat from here to there make one very hungry.

We brought the goodies back, and had a coffee on *Anya*, where Fokke, the Dutchman, was thoroughly enjoying man-talk with Bill. Engine details were swapped, and their little sufferings lovingly discussed. They talked of methods of easing the passage through locks, since the Dutchman was the son of a *batelier*. The women went further into food, and then moved on to poetry, writing and word-processors.

There is a lot to see at Lyon. We were not there long, and had to choose.

The other yachting ladies made a trek to the Lyon branch of Marks and Spencer's, stocking up on tights, pants, English muffins and marmalade.

Later we walked round the lively Place Carnot, bathed in the golden light of the winter sun, went window-shopping along the pedestrianised Rue Victor Hugo; and spent an afternoon at the Musée des Tissus, for the silks of Lyon supplanted those of Italy in the sixteenth century, and today the Lyonnais textiles rival those of Milan.

The mornings tended to be chill and foggy. We watched

in amazement as huge blowers drove slowly, daily, along the quays, piling up the tons of leaves that had descended from the plane trees since yesterday's sweep. The blowers were accompanied by teams of men who bagged the leaves and loaded them into a lorry. The whole procedure held up the traffic and further added to the complications of driving in Lyon.

From the Quai Tilsit, when the mist dispersed, we had a stupendous view from the windows of our barge across to the heights of the old town of Lyon – Lugdunum, the Hill of the Crows, twenty centuries old, towering above the river. Nowadays this hill is called Fourvière, and is crowned by a nineteenth-century basilica whose oddly shaped towers make a useful landmark if one gets lost.

One of the main causes of noise in this part of Lyon is the bridge over the Saône carrying France's main north–south motorway, which in a moment of madness was routed through the middle of Lyon via the tunnel of Fourvière. This is probably the greatest highway-planning mistake in Europe, for to add to the volume of long-distance traffic, which is immense, there are the semi-diurnal ebbs and flows of the local commuters of the third city of France.

Each morning on the national radio the length of the *bouchon* at the tunnel of Fourvière is broadcast in the same way as the weather forecast and the unavoidable jams on the Paris ring road, the Périphérique. If things become really bad a barrier descends at the tunnel entrance, no more vehicles may enter, and streams of cars take the old road – choked, winding and full of hairpin bends. From our berth we were able to contemplate all these jammed-up, hooting vehicles with some equanimity, thinking how nice it was to be travelling by boat. They might, just might, get to the south before we did.

8

Rhône the Rouan

'Mai cuerb lou rose un sagares de neblo: li
couparias em un couteu. Amagon lou
ribeiras, tout, a perdo di visto.'
. . . .
'But a thick fog covers the Rhône:
you could cut it with a knife. It hides
the banks, everything lost to view.'

When one leaves the moorings at Lyon one is not yet in the
Rhône. The confluence of the Saône and the Rhône is to the
south of the city, and as we got under way on a cold and
foggy morning, in mid-November, we still had a kilometre
or two of the Saône to navigate. Under the motorway and
rail bridges we could just make out the river banks through
the fog which reduced visibility below half a kilometre.

We were having a lot of fogs. We have almost forgotten
fogs in England now the cities have been cleaned up.
France has many more than we do, though the French still
think of England as the land of the pea-souper; it is a
reputation hard to lose, since this belief is fortified by films
they see on television of Dickensian or Sherlockian London
wreathed in the sort of fog that now blankets only the
industrial parts of France and the plain of Lombardy in
Italy.

The south side of Lyon is industrial, and the river is
edged with big barge docks, a fuelling station and the
chapel barge, *Le Lien*, which is much used by the barge folk,
who are more than usually devout. Who would not be
devout with the Rhône to voyage on? Even today, after
much taming, it is still to be respected.

Here, at La Mulatière, we ourselves began warily to navigate the Rhône – Rhodanus to the Romans, Lou Rose in Provençale, a rose with deadly thorns, a river whose source in the high Alps is more than a mile above the delta of the Camargue, where it flows into the Mediterranean.

As it descends, fed by torrent and mountain stream, flowing strong and fast, it weaves myth and history, awesome legend and terrible memories of flood and wreck, of torrent and turbulence, of strange water monsters, (the *dracs*, the *lerts*, and the *tarasque*), of giants, of kings, saints, warriors and troubadours. It has been a waterway for two thousand years, this mad bull raging down from the Alps (Rouan in Provençale means both a bull and a mass of moving water, hence it is an appropriate name for the Rhône). It has enticed men to venture, and drowned them, attracted ships with priceless cargoes, and wrecked them, drawn to its banks cities and towns, and inundated them, and still is forgiven and all woe blamed upon its no less rowdy brother, the Mistral.

No sooner in the Rhône than we were into the first of many artificial channels, part of the reconstruction undertaken by the Compagnie Nationale du Rhône (the CNR) since 1941 to provide hydro-electric power and irrigation, and control flooding. This channel avoids the once dreaded rock of Pierre Bénite, and takes one into the lock which now bears its name.

One of the best descriptions of the Rhône navigation in olden days is in *Lou Pouèmo dou Rose*, (*Le Poème du Rhône*) by Frédéric Mistral. It was written at the end of the nineteenth century in the old Provençale language, to the revival of which the author was fervently committed, and which even the French can't read.

Mistral describes the adventures of a group of bargees leaving Lyon to take goods down to the annual fair at Beaucaire on the edge of the delta; how they loaded and equipped themselves for a journey which would take them

away from home for months – three days to get there, and
weeks to struggle home again; how they took on board each
barge the twenty-four horses they would need to pull them
back upstream against the current and up the rapids,
perhaps in the teeth of a Mistral as well.

Their first obstacle was the rock of Pierre Bénite, where
the river flowed fast and dangerously, and where until
recently the awkward shape of the river-bed could cause
flooding to southern Lyon. Now there is a big weir, and the
enormous lock which we entered leads into a lateral canal in
which boats can navigate safely while flood water is di-
verted into the old river-bed. As a result the first ten
kilometres south of Lyon have been turned from what the
guidebook calls a 'disadvantaged and boggy region' into a
vast work-scape of refineries, steelworks, marshalling
yards, tile factories, glass works, and all the other elegant
buildings that make city scenery so pleasant these days.

There is still a strong current in the channel, for after ten
kilometres it returns to the river-bed and we are then
navigating the river itself. Descending the Rhône is like
going down the left-hand side of a liquid escalator, the side
where you don't stand still. In order to keep some sort of
control you have to motor ahead and steer, and you need
control because, apart from the canalised parts, the banks
are inhospitable.

At Vienne the motorway screams alongside the river – six
lanes of frantic, frenetic automobilery which suddenly leap
across from the left bank to the right. In spite of all the
potential navigational dangers on the river, we have an
instinctive fear that we are more likely to be sunk by an
out-of-control truck crashing down through the flimsy rails
of the bridge than from any other cause.

Vienne is largely cut off by pelting traffic, but it has a
good public quay. It was too soon after our start to think of a
night's stop, and we had stayed there before, so we pressed
on, passing on the riverside one of the great restaurants of

France, Point-Pyramide. At one time it rejoiced in three Michelin rosettes, their highest award; then its *patron-chef*, Monsieur Point, died, and such was its reputation that Mr Bibendum suspended for a season his usual ruthless practice of downgrading an establishment on the death of the chef. In the guide of this year it is noted as being closed for rebuilding, and has two rosettes.

The pyramid after which it is named was in mediaeval times reputed to be the tomb of a gentleman known to the French as Ponce Pilate le Procureur, words that to the English ear suggest a satisfactory turpitude. After Pontius Pilate, as we call him, was recalled from Jerusalem he came to Gaul and was given Vienne to govern. Legend tells that in a fit of remorse he threw himself into the Rhône.

We descended down a huge lock bordered by an enormous weir and hydro-electric power station to the old town of Condrieu. This was the great recruiting station for the *mariniers* of the Rhône in days gone by, and is where most of them lived.

Mistral describes the scenes in his *Lou Pouèmo dou Rose* as the boats pause on their journey from Lyon to the great fair at Beaucaire, to bid farewell to the families they would not see for many weeks, and some perhaps never again. Coming from Lyon, where the barges embarked rolls of leather and bales of silk, and having loaded ironmongery at Givors and barrels of beer at Vienne, they anchor at Condrieu. The children have been waiting and watching for hours on the quayside, and the wives on their doorsteps, for the passage of their men. This is how Mistral tells it:

> 'Run! Blandine! your father is coming!'
> 'Throw me the rope, I'll make the boat fast.'
> 'Where is Damian? Here are his shirts.'
> 'What shall we bring you from the fair?'
> 'Tell your Grandpa to buy you a little sister in a pink box!'

'Grandpa, I want to go to Beaucaire.'

'Next time, my boy.'

'Be good, now, have a good time.'

'Keep your spirits up, Auntie.'

'Watch out for the gnomes, the St Elmo's fire and the *Oulurgues*, who haunt the tombs of Arles by night.'

And then goodbye; the Midi draws them, there is no time to lose, says Mistral.

As the Midi drew the *mariniers* of Condrieu, it also drew us, despite the risk of encountering the fearful *Oulurgues* of Arles. After much research and questioning we found that they were spirits who had died suddenly, unshriven, and worse, without revealing under whose mattress they had hidden their hoard of gold.

Nowadays there are still many in Condrieu who work on the big Rhône barges, the *citernes*, which used to be named after the winds of Provence, sonorous and evocative names; but as the craft are now in the hands of a public company they have only dull numbers. Why does the hand of commerce so needlessly have to destroy even the last few vestiges of romance in work?

Condrieu, in company with many other towns on the river, still has its water feasts with the old games of water-jousting, when oarsmen manoeuvred their boats to permit their 'knights' high in the stern to joust against each other, attempting with long lances to 'unhorse' and topple the opponent into the water. There are rowing races and water sports too, and a great atmosphere of festivity, but it all takes place in summer, when the water is friendly to fall into.

We did not fancy a winter swim in the river here, though the sharp and dangerous bend has been by-passed and the ends have been sealed by dams. Part of one dam encloses a pleasure-boat harbour which is too ritzy and expensive to be a good transit stop for a barge, but most of the disused bend is sealed off and, in the English words of the trilingual river guide, 'has a sheet of water which is exclusively supplied by clear infiltration that invites for swimming and wind-surfing on a supervised shore'. Well, this is fine for summer swimmers and windsurfers, but it was no use to us in winter.

The weather had now turned really cold, and in eager need of the sun we moved further on and stopped overnight at a disused commercial quay at St-Vallier.

There was no sun. It was pouring with rain. From the little boat already moored ahead came a Cape Cod 'Hallao!' and up popped the unmistakable yellow curls, with peaked cap rakishly perched on top, of Ding Watson, who bravely helped us with our lines.

Imagine something a bit like a wooden poster hoarding twenty feet high. Throw Rhône water at it, blow it dry with Mistral, and splinter it with careless barges. Decorate the bottom few feet with slimy green weed, and provide very few footholds, and you will have the mooring at St-Vallier. So Ding's feat of mountaineering in the rain was beyond the call of friendship, and much appreciated.

Shortly after St-Vallier one passes the lock at Gervans, and once again enters wine country, more Côtes du Rhône, this time the Hermitages, a renowned wine said to have a perfume of raspberries. One also comes across a rock which most *mariniers* would have preferred to have blown up as a danger to navigation, but pseudo-nostalgia has prevailed upon the authorities to keep it intact. It is called the Table du Roi, and legend has it that King Louis IX, later to become St Louis, had his dinner on it when descending the Rhône on his way to the Crusades.

If he did he must have been *dingue* (daft) or drunk. It is a small flat rock a metre or two square, and less than half a metre above the normal water level. The current is still strong; in those days it was almost certainly stronger, and the problems of keeping the royal barge safely moored, while the King feasted with the one or two others who might have crowded on to the rock without nudging the royal elbow into the royal soup, would have been formidable.

The King's Table is still a notorious hazard to navigation when conditions are adverse, for the current flows slightly across the channel; the water level can rise to cover the rock, and debris occasionally sweeps away the flimsy pillar which marks it. The long barge trains do not like it a bit.

A little further on we passed the town of Glun, now somewhat isolated in the royal river on the right bank. This was the bank known to the old boatmen as '*Royaume*', ('Kingdom') referring to the Kingdom of France. The other, the east bank, was '*Empire*', belonging to that usually considered to be neither Holy, nor Roman, nor an Empire. Thus the old *mariniers* would call out '*Empire!*' or '*Royaume!*' according to which bank they wished the helmsman to steer for. To travel south to the Mediterranean, you keep 'Kingdom' on the right.

Glun is famous for its *friture du Rhône*, or fry-up of small fresh-water fish. One finds all over France *soupes de poissons*, stews and *fritures*. These are infinitely variable fish dishes, some of sea fish, some of river fish – in effect all that is locally available in the way of small fish thriftily cooked together – under names such as *Pocheuse* (on the Saône), *Cotriade* (in Brittany), *Caudière* (in Picardie), and *Matelotes*, which you find almost everywhere.

The Marseillaise dish of *bouillabaisse*, perhaps the most famous of this genre, must have started out as a fairly simple combination of whatever fish were handy, with the addition of a prawn or two, put into a savorous Provençale broth of oil, garlic and tomatoes. Tradition says that Venus invented this fish stew, which she flavoured with saffron, known in those times as a soporific, to put her spouse Vulcan to sleep for her own devious ends. We do not observe Frenchmen eyeing their wives with deep thought when given this now very grand and expensive dish. It is not something to eat alone – there must be at least two of you or the restaurant thinks it is not worth the bother. It contains the most expensive fish, plus an obligatory *langouste* or two: and there is more mystique and mumbo-jumbo about it than Macbeth's Weird Sisters could conjure up.

The giant Gargantua ate his own special kind of *bouillabaisse* at Glun; being thirsty he bestrode the Rhône, and scooped up in his hands a large amount of the river, swallowing boats, men, horses and all. He may have thought, as many do: too many bones. In which case, one forgets the *bouillabaisse* and tries the excellent *soupe de poissons* that is served along the coast of Languedoc; it is pungent, aromatic and sieved – no bones. One can, however, make an excellent sardine *bouillabaisse* at home, without breaking the bank.

Back to the Rhône. From here on the way downriver is fast and brutal (we are entering the part of France that plays Rugby football). Broad-flowing river, hydro-electric stations, nuclear-power stations, enormous weirs and huge locks with flashing lights controlled from high towers.

Now instead of the lock gates being in pairs, and closing on hinges like doors, we encountered something ominously called a guillotine which rose from the water ahead of us, dark and dripping, with just enough clearance for us to get under. A steady rain of heavy drops swept and drummed along the boat as we passed, sending our little kitten scurrying indignantly and damply for cover, for she enjoyed the bustle of locking. She had become more careful not to get in the way, and we felt able to give her the freedom of the boat.

Quays along here are either reserved for the *commerce*, and they mean it, or are unapproachable. Some in the guide are labelled *délicat*; in fact they are usually very robust and would soon wreck a delicate boat.

We passed broken quay after sloping quay, declining all, until finally at nightfall we had to make fast to a quay that sloped back so alarmingly that it must have been designed by a madman. It is close to the huge nuclear-power station of Cruas, which has a lovely quay to which ordinary non-radioactive folk like us are not allowed access. Fortified by the unlikelihood of another nuclear disaster so soon after the Russian one (there is nothing like a nasty accident to make people take extra care), we spent a good night, again in the company of *Anya*, even though the wind was rising from the north-west, which, though favourable to our progress, was not a good portent in these parts.

Weather omens notwithstanding, the next day broke fine and clear with only a moderate wind, and we made good progress. In the canals a day's run of thirty kilometres had been good going, but now we could easily chalk up eighty-odd kilometres per day since the speed limit is too high to concern us (35 k.p.h.) and locks are few. We did not travel at our full speed, as this burns up fuel too quickly; the curve of fuel consumption against speed is exponential – the last knot or two is achieved at disproportionate expense. We went through the water economically at about ten or eleven kilometres per hour, and our rate of advance down the big rivers varied from that to a maximum of about fifteen, depending on daughters and currents.

But this day was not to be a joyous one. We passed through the spectacular scenery of the Défilé de Donzère, where the Rhône cuts through a steep gorge. On our left were immense dark cliffs towering above the river, and the little yacht *Anya* under them caught the sunlight and looked like a toy boat. We passed the uranium refinery near Pierrelatte, but could not see, as we had hoped to, the huge rock that gives the town its name; it was shaken out of his shoe by the giant Gargantua.

He would have felt at home with the impressive Bollène lock which we were now entering (probably using it as a hip bath). At the time of its construction it was the highest in the world; the water level drops twenty-six metres, and at its lowest gives one the impression of being in a vast, dark, roofless cathedral after a downpour.

It would be impossible to make a boat fast to fixed bollards when the water level changes more than the length of a cricket pitch, and the lock has been designed with large bollards on platforms that float up or down with the boats.

While we were photographing the descent, the kitten Tassie must have jumped on to one of these floating bollards; we heard her calling but could not find her, and she was never seen again. Presumably she fell in, for there was nowhere to climb to except back on board, and if she missed her footing she would have stood no chance in the raging whirlpools of turbulent water, for the lock is drained through large ducts built deep down in the floor.

The lock-keeper, on the radio, agreed to hold things up for a moment so we could search, but there was no sign and no sound. We had to face facts and move on. There was no time or inclination for blame; who should have done what, or didn't, was immaterial. After having a ship's cat which had lived in our boat for more than ten years, the mortality of cats on board had suddenly become alarming.

We had not become as intimate with the little stray as we had been with Twosie.

The latter had been an affectionate and well-mannered cat, well adapted to living with humans, while Tassie had remained untamed and had held us at arm's length, except for the violent games she liked to play with Bill, whom she loved in her own wild fashion. She had come to a sort of truce with us who fed and housed her, but it was an armed truce; she resented strangers on board, and we half expected her to leave us one day to go her own way among woods and fields, but not like this. It was a different sort of grief that we felt. Often grief is a question of coping with what the griever has himself lost. In this case it was mainly sorrow for the little scrap who had come to such an end.

We moored at Roquemaure feeling heavy. The weather remained bright and fine, the wind moderate north-west. We had turned the boat to face into the wind and current in order to moor, and our bows were now facing north-west. It was a good quay, of stone and covered with stone chippings on a *levée*, some distance behind which sheltered the mediaeval city crowned by a huge castle, the Rocque Maure, or Saracen Castle.

Anya berthed ahead of us. They had been with us in the lock at Bollène, and had helped us in our search, and now came to offer sympathy. We joined together in a walk round the village (for with 2300 odd inhabitants it is no longer a city), doing some shopping and finding somewhere to eat out; Laurel didn't feel like cooking.

9

Mistral

'Boufo, bregand de Coucho-Mousco!
Boufo, Desbadarna de Dieu, que te
crebesses! l'aura dounc jamai res, o
Manjo-fango, que tapara lou trau de
mounte sortes?'
. . . .
'Blow, you Fly-swatting Brigand!
Blow till you burst, you Godforsaken
Draggletail! Is there no one,
Soil-swallower, who will come to
stop up the hole you blow from?

Our last few moorings had been impractical for shoregoing, so this was our first walk ashore since Lyon, and suddenly we realised we were in Provence at last, for here on the outskirts of the village we passed Les Arènes, the Bullring, looking impossibly rickety, the bright red paint, faded to a Tyrian rose, peeling off the wooden barricades after the summer's sun. The leaves of autumn eddied on the sand. One could almost hear the wind in the plane trees roaring 'Olé!'

The town square was also guarded by the usual plane trees, huddling together and holding gnarled knuckles to heaven, praying for the spring. They were ancient, and many of them were hollow. The hollows were filled with big stones. 'They weight them down against the Mistral,' said Laurel gravely. We found no one to tell us if this was true.

The four of us found a good meal at Le Pressoir, another welcome reminder of Provence, the old olive press still in evidence among the trees in the courtyard, where the tables would be set in the shade in summer. Laurel had looked forward to the first olive tree, an important event, she felt, meriting a celebration. We felt our spirits rise as we entered a cheerful barn of a place, with red checked tablecloths and wholesome earthenware jugs of wine, where the staff and the ambience were both friendly. Lamb cutlets were grilled over charcoal, we talked, downed a second jugful and felt much better. We ought, in retrospect, to have paid more attention to the sky as we walked back to our boats, instead of marvelling at the nearly full moon and the occasional racing clouds that passed in front of it, but our minds were full of interesting topics and our stomachs full of lamb.

French lamb is different from British and New Zealand lamb, and the French are quite firm about preferring the taste of their own meat, while Bill in particular prefers the stronger taste of the English type, finding the young French lamb rather insipid. They do tend to eat their lamb younger than we do, and Bill believes they eat it before the taste has set. Certainly the French *mouton*, which we would call last year's lamb, has a lot more flavour and is more to our taste. The French are astonished at the price at which British sheep farmers can produce lamb, which in France is very expensive. The problem seems to be that most of the French mountain sheep farmers are in a very small way of business; it is *artisanal* rather than mass-production, hand-raising rather than business-farming.

By the time we got back aboard the wind was piping up a bit, but it was still from right ahead, the sky was clear, and the moon shone bright, so we turned in feeling easy.

By two in the morning our friends in *Anya* were up again, and so was Bill. There is an agreement that he gets up first at night to check anything, and calls Laurel if help is needed, which means that she becomes far more anxious waiting for

the call than she would have done if she had got up in the
first place. However, it does give her time to sit on her socks
and warm them.

The wind was up to gale force and had veered round to
the north, so it was blowing us on to the quay at an angle of
about forty-five degrees. The waves that were forming over
a fetch of about a kilometre, though not large, were enough
to bounce the much smaller yacht *Anya* against the stone
quay, and they were having a very rough time. After a brief
chat about it, shouting above the wind, they decided to go,
and cast off heading south to look for a more comfortable
spot.

They were wise. If we had known then, as we were told
by a *marinier* later, that the villagers were rumoured to push
their old cars over this quay, and at low water boats had
been known to ground on them, we might have left at the
same time and saved ourselves some anguish.

As it was, *Hosanna*, a much bigger and very robust boat,
was not much worried by the waves; we adjusted our
tyres and went to bed again. When people say, 'How lovely
to live on a boat all the time,' and we reply, 'Well, ummmm
. . .' it is nights like this that we are remembering. They
balance the times when we sit on our boat in warm sunshine, looking at a millionaire's view with a tinkling glass of
something cold in our hands, in perfect peace. One gets
nothing without paying.

We were up twice more to adjust the tyres, and by 5 a.m.
a full Mistral was blowing, the legendary (but all too real)
wind from the north that comes from the snows of the
Massif Central, and sweeps down the Rhône valley, collecting speed and energy as it goes. A whirling, drying dynamo,
evaporating, freezing, turning puddles to clouds of dust,
bending the trees double, slowing down cars and trains,
and building up great waves and hissing spray on the
waters of the Rhône, before roaring down to the low-
pressure areas over the Mediterranean – this is the Mistral.

'*Maître*' is one of its names in Provençale, though there are much more uncomplimentary ones.

We rose at daybreak after a virtually sleepless night. We had a miserably uncomfortable day, during which Bill walked into town and came back with three more tyres, but with the forecast of a lessening wind we decided to put our feet up and ride it out. There followed another atrocious night of judder and cacophony.

Next morning, instead of lessening, the wind had increased and the temperature had dropped to about freezing point. Four of our six tyres had broken adrift in the night, and it was now so rough that spray was continually breaking over the boat and we were covered from end to end with a coat of ice, almost two inches thick in places. The sky was so intensely blue, and the sun so aggressively bright, that feeling numb with cold seemed all wrong. Was this the Provençale sunshine that we had sought for so long, and was it always accompanied by this wind shrieking like the souls of the damned, and this perishing cold?

Where the spray blew over the quay and coated the pebbles they had frozen too, making the stony surface flash like diamonds in the sunlight, and nearly impossible to walk upon.

By now *Hosanna* was crashing against the quay, despite our efforts to moor her firmly and cushion her with tyres, and with every judder something fell off a shelf down below, as we were not stowed for sea.

We tried to spring the boat off the quay and leave, as *Anya* had done, but we could not get her off; she was now tightly pinned by the wind, so, freezing, tired and hoarse with shouting above the noise, we abandoned the attempt. All the time the violence of the wind was increasing and we were being bounced so hard against the stone quay that the vibration shattered a light bulb. It was time to think positively.

Bill stayed with the boat in case of 'something worse' and

Laurel walked into the town in search of more tyres; we were losing them at the rate of one every hour, which was the length of time it took for the ropes holding them to chafe through.

A lady with a dislocated hip cannot be expected to carry, or even roll, several tyres back over about a quarter of a mile, but Laurel is a resourceful lass. It was an eventful walk. With the wind shrieking in her ears she started with a difficult traverse of the iced pebbles on the quay, landing on her *derrière* at least once, and in no state to admire the undoubted beauty of the plants and grasses, stiff and shimmering with ice, everything hard and crystalline. Walking against the wind was difficult; the thump and buffet of the gusts was like being hit with a full kitbag.

Arriving in the village, she prudently bought bread and then commandeered the first car she saw; it contained a completely astonished young Frenchman on his way home to lunch, who drove her to the garage and brought her back to the quay with as many old tyres as she could persuade him to get into the boot of his car. He even staggered across the icy pebbles and helped us get them to the boat. It was a busy scene for a time as the boat was taking a lot of punishment, and all attention was focused on getting the tyres in place.

As usual, the people who lived in the village had no idea what was happening down on the Rhône, and our benefactor was amazed at the violence of the wind. We hope we made our eternal gratitude clear to our unknown helper, who understood the situation sufficiently to return with a second load of tyres. We yelled our thanks as he disappeared down the *levée* with a story to tell over his belated lunch. ('Down on the Rhône, it whistles a full Mistral . . .')

Now that we had more tyres, arranged two thick, the ride became tolerable for a time. We had a rotten lunch, as we were cold, uncomfortable and worried about our boat. Two dreadful events in a week, had we already had the third of

the series or was it still to come? Bill is not exactly super-
stitious, but it doesn't do to fly in the face of providence.

Then the wind grew even stronger. It stopped howling
and started screaming. It was impossible to stand on the
quay with nothing to hold on to, and Laurel was literally
blown off her feet. The tyres were not coping, we were in
danger of sustaining real damage, and to save our home we
had to face the prospect of trying once more to get away
from the quay in worse conditions than had previously
made us fail.

Clearly it was no good doing the same thing over again,
and Bill went out to study the problem, sitting in the icy
wind for about ten minutes, wincing every time his baby
banged against the stone wall. Laurel also winced as yet
another row of books fell off a shelf.

The quay, which was about forty metres long, was built
out from the high grassy bank, and its corners, or knuckles,
extended about five metres into the stream. It was about
three metres deep alongside the quay, but the bottom

would shelve back to the normal river bank at either end. We moved *Hosanna* back until her stern overlapped the downwind knuckle by about eight metres, tied our oldest rope from the stern to a ring on the knuckle, and left Laurel holding on to our end, where it was safely turned several times round the ship's after bitts. Bill cast off all the other ropes. The wind held us pinned. The wet rope froze into a solid bar.

With the engines going astern at absolute maximum power the rope tightened until it was humming. As it stretched the friction melted the ice in it. After what seemed an age, with the hooligan wind raging to hold us against the quay, our bows gradually started to turn into the wind as the boat became a lever with the knuckle of the quay acting as a fulcrum. As the bows went out, so the stern, with its propellers underneath it, moved into shallower water. Would we get head to wind before the propellers bit into the mud, stone and detritus that lined the bank? If that happened, our situation would be desperate.

It took more than a quarter of an hour, gaining a little in the lulls, losing a little in the gusts; the engines screaming as loud as the wind, the spray that swept over the boat turning in the icy air to frozen pellets that rattled as they hit the steelwork and bruised the flesh as they hit our faces. As the bows came at last head to wind, Laurel waited for Bill's signal to let go the rope that had been preventing us from going backwards. The moment had to be precisely judged. 'Let go!' he shouted, with a thumbs-up sign to make all clear.

We jettisoned the rope, throwing it clear. It didn't seem a good idea to have a lame lady trying to haul inboard fifty feet of heavy, freezing rope as thick as a man's wrist, while trying to keep her feet on a deck covered in ice. She is not a bad hand with a rope, but this was tricky. She is too valuable an asset to leave behind attached to a mediocre rope. We sacrificed the rope.

As *Hosanna* turned downwind, free of the shore, some of the bite went out of the Mistral. Everything became a bit easier as we rounded a bend, and the noise grew less. It even felt warmer, with the sun coming through the wheelhouse windows as we turned towards the south again.

It was not far to Avignon lock, where there is a large hydro-electric plant. We locked through after an anxious half hour waiting for a huge upstream barge to clear – anxious because not only was there a violent wind blowing us towards the hydro-electric plant, but there was a strong current pulling us as well. We waited about two kilometres upstream; this would have been a bad time to have an engine failure, and we needed room to get an anchor out if the need arose. Not that Bill had a lot of faith in the holding power of an anchor on what was almost certainly a rocky bottom; it might be good for the taste of pike and *sandre*, but it is the worst holding ground there is.

On the lower, leeward side of the lock there were several dolphins, or mooring posts, to which we could make fast in the wind shadow of the huge walls of the hydro-electric plant, in company with another small barge bound upstream waiting for better weather. The posts stood out in the stream, with access to the shore only for an agile chimpanzee, but we were moored in a safe and sheltered spot.

The relief was unspeakable. We watched the trees on top of the bank, some forty feet above us, bending double like gymnasts limbering up; we stopped shouting at each other, put the books back on the shelves, and, if gusts occasionally moaned down our chimney and made the stove give off a puff of smoke, it was a small price to pay for the comforting warmth it gave.

We consoled ourselves with a batch of Scotch pancakes and syrup. It was marvellous to revel in peace and comparative tranquillity, and have the heavy weight of anxiety lifted off us.

We were even able to regard with equanimity the 'PHEN' mark painted on the wall opposite us. This stands for *'plus hautes eaux navigables'*, the highest point at which the water is considered safe to navigate in times of flooding. We marvelled, as the mark was probably a good twenty feet above us, and found it hard to imagine floods of that order. Laurel, with logic, worried that precisely when you needed to see the mark it would be covered with floodwater, but at such times, we found, warnings would be on view at every lock.

Gradually the enormous icicles that had formed on our masts and rigging melted in the sun, for the weather had remained bright and the air was dry; our icing-up had been caused by the super-cooling effect of the high wind.

We had one more day like that, then the wind eased. Tradition says that the Mistral lasts three, six or nine days; ours lasted four.

You now have some idea of the depth of meaning in our earlier warning that it isn't always a piece of cake on the inland-water route. We were unwise to have stayed in the berth at Roquemaure after *Anya* left. Conditions were already intolerable for them; they were not so for us and the weather forecast was for an improvement. Still, our bad decision, however well founded, could have wrecked a flimsier ship.

We arrived at the city of Avignon at eleven in the morning, passing the famous half bridge (which is all that remains) of St-Bénézet, on which one used to dance in a round, and still can on payment of a modest entry fee. (It's a little difficult to envisage dancing *tous en rond* on a bridge so narrow, but no doubt they managed.)

During the Middle Ages there were only three bridges over the Rhône between Lyon and the sea: at Vienne, at Pont St-Esprit (another old bargemen's town) and at Avignon. The last two were completed at the beginning of the thirteenth century by the Frères Pontifes, founded by St

Bénézet, who, while tending his flocks in the Vivarais, was commissioned by an angel to build a bridge at Avignon. The city authorities found this just as hard to believe as they would today. 'You want to speak to the Sire? Your name? Bénézet. Profession? Shepherd. Business? Permission to build a bridge. Over the Rhône. I see. On whose instructions? What was that again? An *angel*? Just a moment while I consult my colleague . . . No, sorry, the Sire is in conference and cannot be disturbed.'

In order to convince them St Bénézet had to start building his bridge alone, lifting massive rocks with such ease that it was clear to all that his angel must be eagerly assisting in the work. From then on money and help flowed in and enabled him to found the Bridge Brothers, and after they had built the bridge at Avignon in a mere eleven years they went on to build Pont St-Esprit, with a yet more prestigious entrepreneur this time, the Holy Spirit Itself.

After the terrible weather we had had we were only half surprised to find the *port de plaisance* in Avignon wrecked – though not by the recent Mistral. Some weeks previously the whole of France had been shocked when the old and beautiful city of Nîmes had been inundated by a ferocious and sudden flood. The cause of this disaster had been a terrible cloudburst in the mountains just above Nîmes; the water (as much as falls on Paris in a year) had deluged the streets up to three metres deep, bearing away and wrecking cars and killing nine people.

Similar storms had occurred on the high land to the north-east of Avignon, up in the Massif de Ventoux (The Windy Mountain). The new flood-control works on the Rhône had coped with the water that came down, but the river had swelled suddenly. By the quayside at Avignon it had risen three metres in an hour and had flowed faster than a man could run. All the elegant new pontoons the local Chamber of Commerce had placed to moor smallish boats (up to about twelve metres) had been bent, twisted out of shape or just washed away by the fast-flowing water. Boats that had survived were now moored at the stone quay, normally the place for barges.

The pontoons had been made of aluminium. The French place a lot of faith in aluminium, as they invented the process. In the last century they originated the first commercial smelting in the world with a newly discovered ore, called bauxite after the place in Provence called Les Baux where it was mined. For boat-mooring pontoons it is not a strong enough material to take the sort of battering that steel will take in its stride.

The Avignon port would have been all right if the pontoons had been made of steel. Rusty perhaps, but still there. As it was, nothing was being done to clear up the mess, because the three parties to the contract – designer, builder and insurer – were arguing about who was to blame. A story we have heard before in other parts of the world.

Meanwhile, we had nowhere to berth, for the quayside area meant for larger pleasure craft was full of small boats which had been rescued from the pontoon debacle. We found ourselves a spot that said 'No mooring for pleasure boats' and made fast. We were still tired, and patience was not all it might have been.

It appeared that the harbour master was away on leave, probably needing a good rest. His Belgian deputy came to tell us that we were illegally moored in a place reserved for the *commerce*, and Bill replied that as there was no other place to moor we were staying, but would gladly offer our place to a commercial barge if one would appear. The Belgian deputy rocked from foot to foot in his embarrassment. We would lose him his job. Bill observed that if he was any good at it he would be very unlikely to be sacked for a mild misdemeanour that was after all outside his control. It was we, was it not, who were committing the crime. Come in and have a drink.

The deputy's worries seemed to fade into the background and, consoled by the liquid lubricant of amicability, he left to return to his office. Shortly afterwards four more boats, two of them known to us from our time in Lyon, appeared and moored close to us. *Anya* was already there.

The deputy appeared once more to wring his hands and do his little dance, so we had to give him another transfusion of confidence. There is a well-known axiom among boat-wandering people: when the situation becomes awkward, act wet, apologise afterwards and be generous with the bottle.

The tranfusion offered to the harassed official gathered other folk with some sort of interest in boats (and perhaps several without any at all), and it turned into an informal party.

The next day Bill went with Fokke, the Dutchman, to Lyon to collect the two motor cars with which we were now encumbered. (We had been unable to get the little one back on board owing to a shortage of Belgians.) We took parties of fellow travellers out to the hypermarket on the outskirts of town, and then, as we were planning to go on soon, Bill took the little car and parked it on the deserted quay at Arles.

While the men were away Laurel and Ding went to market, always an enjoyable experience, and then had a discussion over coffee on the subject of 'lack of communication capability of the male in times of stress'. They found they had a lot to say about it. Laurel felt that the subject of verbal communication should be taught, along with navigation and seamanship, so that she might not then find herself, in the middle of a Mistral, at the wrong end of an indecipherable bellow to 'Get that damn thing there, quick, and tie it on to here. NO! HERE!' She had the advantage of knowing what most things on a boat were called, if only Bill could come up with the right name in moments of difficulty. Everyone in the family knows that if Bill asks for a screwdriver at mealtimes what he really wants is a corkscrew – nobody has to think about it. But yelling for the marmalade in a Mistral when you actually want the marlinespike is confusing. We call it translexia. Women can't cope with it. On the other hand Bill thinks Laurel's brain goes soft when threatened with high winds, clearly a case of pre-Mistral tension. Both problems seem to get worse with age.

Ding had suffered. Not only did she not yet know all the names of the bits of boat, but when Fokke got agitated he forgot to speak English and gave loud and urgent instructions in Dutch, which she neither speaks nor understands. The subject was absorbing, they drank a lot of coffee, cried on each other's shoulders, swapped recipes and had a thoroughly good time.

There was another convivial party that evening, as everyone had now recovered from the Mistral. Like the Blitz, everyone had their Mistral story. *Dolfijn* and *Lambrusco* had been safe in the *port de plaisance* at Condrieu. *Anya* had come to Avignon and found a haven among the twisted pontoons, though Ding and Fokke were harassed daily by the deputy harbour master, who wished them elsewhere.

A bad Mistral is known to do all sorts of things to people and the country; it is inextricably tied up with myth and legend, so great is its effect on the life of the Provençales. It drives men and animals mad for a start (its effect on women is not recorded but Bill has a few suggestions). It eats earth, being known as Le Mange-Fange, which is a recognition of its habit of blowing the topsoil off the fields and taking it off towards Africa. On all the arable land in the valley one sees hedgerows and thick lines of cypresses running east–west to protect the crops from the wind's ravages. It is feared as few other winds in the world are feared, because the full force of the gale is funnelled down the narrow gorge of the Rhône, and its normal velocity can be doubled in certain places.

Later, with our own spirits restored (but our bottled ones somewhat reduced), our cosmopolitan party of English, Dutch, American, German, Scandinavian, Belgian and French went ashore to eat in the Greek taverna.

In most countries of the world, and Britain and America especially, one finds restaurants of many diverse types. One can in a small town like Great Yarmouth enjoy the *cuisines* of Greece, Cyprus, Turkey, Pakistan, India, Indo-China, China, Malaya, Polynesia, Mexico, America, France, Italy and Germany. And fish and chips too. But even in a big city in France it is rare to find any other sort of cooking than French and Indo-Chinese. Sometimes recently there is a McDonald's, if you can call that *cuisine*. Here in Avignon was a Greek restaurant.

We were a little noisy; it was late, and it had been a good party. We didn't expect any dancing, and we didn't envisage throwing the crockery about, but as in most tavernas we had hoped for a few Greek songs: 'Varka sto gyalo' and

'Strosse to stroma sou', perhaps. But the atmosphere was quiet and reverential. It was in fact a typical French restaurant serving Greek dishes – nice ones, but we are used to more noise with our Mezedes and Moussaka.

It is a fact of French life that their better restaurants always have a reverential air. Diners speak in hushed tones as if in church, and there is a distinct sensation of worship, a tinkle of Holy Water. Clearly the French upper strata believe that noisy enjoyment impedes the flow of gastric juices and curdles their beautiful sauces to vinegar and ashes. Or perhaps it is a question of manners? Does not every well-bred child in France grow up to the refrain of his mother's words: *'Tais-toi et mange'* ('Be quiet and eat')? Sometimes they dine as if doing penance for some unknown sin, atoning before the altar of gastronomy. Lively conversation and laughter were to be found in the village café-restaurants, and even some of the little red 'R' establishments, where we would often be drawn into whatever festivity was in progress, and no chef could doubt that the food was appreciated. In the classier places, a loud burst of laughter turns heads, as if one had applauded between movements in a symphony concert. On the other hand, France does have the best restaurants in the world, so perhaps hushed appreciation is appropriate.

For those wishing to try more traditional *cuisine* in Avignon, we can recommend La Fourchette in the Rue Racine. This excellent and inexpensive restaurant has been given Mr Bibendum's little red 'R', which we find a reliable indication. It is very tiny, however, crowding six tables and a pot plant into the hallway, so if you are going for lunch, either reserve or go early to beat the local businessmen.

The deputy harbour master was moaning again the next morning. No *commerce* had appeared wanting to berth at the town quay, but he seemed to be on the point of bursting into tears at the slings and arrows of his outrageous fortune in working for the Chamber of Commerce.

Bill observed that if the Chamber of Commerce wanted tourists they should do something about encouraging them to stay, like getting the port repaired, instead of making their lives a misery. He said that he had had enough of the port of Avignon, even though the city is undoubtedly one of the most beautiful in France.

He took the address of the President of the Chamber of Commerce, intending to write him a note extolling the virtues of flexibility in dealing with human problems, but he never got around to it. He seldom does. He spends a half-hour calming himself by drafting acidly worded polemics (no communication trouble at all on paper), and then forgets to send them.

So we set off again and passed between the twin towns of Beaucaire and Tarascon, under the loom of the great stone *châteaux* that stand on either bank, thankful that the fearful *tarasque*, or dragon, that lived here in the Rhône, once a devourer of unwary *mariniers* and washerwomen, is no more.

St Martha, newly arrived from Stes-Maries-de-la-Mer in the first century AD, marched to meet the beast, sprinkled it with Holy Water and offered it a cross. The beast was instantly calmed, she encircled it with her girdle, and led it to the people of Tarascon. Who stoned it to death.

Unfortunately there is no place to stop on the Rhône at either of these lovely old towns, and one would like to suggest to the tourist authorities at both that they would have a lot to gain by following the examples of Arles and Avignon.

Beaucaire has a large basin of moorings, but access to it from the Rhône has been cut off by the changes in the navigation, and to get there one must go a long way round via St-Gilles.

As we have said, the town of Beaucaire held one of the biggest fairs in Europe on the banks of the Rhône, and it was to this fair that the boatmen in the *Poème du Rhône* were going. Wares came from all over Europe and North Africa to be sold at the fair; and the whole town was full to bursting with traders and merchandise, fairgoers and entertainers. The names of its streets (where they have not been renamed after politicians) recall the commodities traded.

The coming of the railways caused the decline of the fair, which was essentially an annual event based on the critical time when ships could arrive from Italy and North Africa in good summer weather, and barges could descend the Rhône and have a chance to make the long re-ascent before the autumn rains started to flood.

Now all that remains is an annual *fête*, to which come the acrobats, entertainers and roundabouts that were once themselves the side-shows of the commercial fair. And of course the *Tauromachie*, for Beaucaire is one of the traditional bull-fighting towns of the Midi.

We had to pass by, dodging the new bridge which is a long time a-building – no angels to assist, evidently. We forked left at the dividing of the river, where the Petit Rhône branches off to form the western arm of the delta, and turned the corner into Arles.

10

Journey's end

'Eilalin a la baisso, i canau de Beu-caire e
d'Aigo Morto, per carga li bladeto de
Toulouso, li vin dou lengado, la sau de
muro.'
. . . .
'Down in the Midi, in the canals of
Beaucaire and Aigues-Mortes,
loading grain from Toulouse, the
wine of Languedoc, and salt from the
sea.'

At Arles we had the third of our misfortunes. As we
rounded the bend to approach the old high Quai Lamartine
on the left bank, ('Empire'), we saw our little car upended
on its nose in the shallow water at the foot of the quay.
It had been well parked right away from the edge and
parallel to it, and could not possibly have rolled there,
and with a lot of experience of parking on quays we
know how to do it. Had the dreaded *Oulurgues* been at
work?

We drove *Hosanna* on to the new mooring pontoons on
the 'Kingdom' side, and were shouted at by a French
yachtsman, well established there and living aboard his
boat, to the effect that we were too big and should go
somewhere else. We were in no mood for this sort of thing,
so we ignored him, especially as a fellow English lady
invited us to berth alongside her boat. There was exactly a
Hosanna's length available, and in spite of a strong current
Bill slotted the boat in as if she had been a Mini, without
touching the boats ahead or astern. Laurel was proud

of him, and hoped that unwelcoming Frenchman was chastened by this superb piece of boathandling.

Pleasure boats at Arles may berth alongside this large pontoon, which the town's Chamber of Commerce has thoughtfully provided and equipped extremely well. To encourage visitors they provide not only free moorings, but also free water and electricity, an almost unheard-of concession, not only in France, but anywhere we have been. A problem arises because the word quickly goes round and the water gypsies move in and soon occupy all the space. To live rent-free, with free water and electricity, is everybody's dream.

This was the situation at Arles, where there was no doubt that the pontoon was overcrowded with long-term free-loaders, including Germans who had left their boats there and gone home for the winter; and not the short stay tourists for whom it was intended. (Alas, this municipal generosity must be doomed if it is so abused.)

The nervousness of some occupants was a result of the flood that had caused all the damage at Avignon. The water, swollen still further by the time it reached Arles, had risen six metres above normal, but the fire brigade had turned out to see to the safety of the mooring pontoon and it had withstood the strain, albeit with some minor damage. It had been a thoroughly frightening experience for those who had been on boats there at the time. The Rhône is not a gentleman's river, it is a hooligan of a river.

Talking of hooligans, Bill went over to the Quai Lamartine and looked at the little car. Its wheel tracks were clearly visible in the mud, and so were the footmarks of a number of people. It had been pushed back and forth, had been turned and deliberately pushed over the edge into the water some six metres below. (Quays and river banks at Arles are extremely high because of the risk of flooding.)

The little car was old, she had run all over Europe for us for years, spares could only be obtained in junk-yards, and she had been showing signs of needing a new clutch soon. On the other hand she was a good goer, and was the smallest car obtainable, which makes for convenience in boats. Now she was clearly a wreck. No car drops twenty feet on to its nose and comes up smiling.

Bill pondered a moment. The chance of finding the hooligans was remote; even if one did, any hope of obtaining damages was equally remote, and the hassle of fighting for them would be great, tying us up for ages. The sum involved would be small, the only motivation for pursuing the case would be revenge, and revenge is a luxury that one can manage without. Just.

There was also the canny thought that we might well be liable for the cost of removing the car from its resting place at the foot of the quay; such injustices are not unknown in this unkind world, and that would be the final insult. The aggravation would be insupportable.

The car had Dutch number plates. He walked away from it. 'Our car? No, very sure, it is Dutch, see, and we are English.' Bill felt a bit like St Peter, and expected to hear a cock crow at any moment. We watched from our berth as the sad little car was removed by crane the next day. There was a small crowd. It is the Arabs, they said, always the Arabs.

There is a large settlement of Algerians in Arles, and unemployment among them is high, which exacerbates France's principal racial problem. The young unemployed have to amuse themselves, and doing damage is as good a way as any.

On the road to the railway station nearby we saw one day that the windows of every parked car had been smashed in. This is not a big city. Arles has a population of 50,000. Not forgetting the *Oulurgues*.

Hooliganism in France is often given a healthy outlet, in the South at least, at local festivals where the young can prove their manhood with the bulls under the eyes of their elders and girlfriends. One has only to be on site for a *fête votive* in the bull-fighting areas of the Midi (of which Arles is one of the principal) to see scenes when the bulls are driven through the streets that would make the average football fan's hair stand on end.

Bill observed that the hooliganism at Arles was all Vincent Van Gogh's fault: he had started the rot by cutting off the lobe of his ear with a razor, which gave the locals a taste for blood, and after that it has all been downhill.

There was at the time a major exhibition of Van Gogh's works assembled from various parts of the world, it being the centenary of his arrival in Arles. He spent almost two years there and was unwell, ill-nourished, unkempt and badly treated by the inhabitants (except for the postman, one of the few who befriended him). He painted like a man possessed, as indeed he was. In 1889 a petition signed by thirty Arlésiens put him into a cell at the Hôtel Dieu, then the asylum. It seems that the tormented Dutchman has now been forgiven, and the Provençale town has absorbed him into their culture.

The exhibition was ironically sheltered in an art gallery in the cloisters of that very same Hôtel Dieu, renamed l'Espace Van Gogh. The security necessary to mount an exhibition of works of this value is so intense that one's enjoyment of the pictures is affected.

In this case there seemed almost to be a security man for each picture. There was a barrier erected so far from the paintings that it was quite impossible to see any detail, and when Bill, who is a tall man, leant across to get a closer look at a drawing he set bells ringing. A discreet hullabaloo ensued, fearfully quiet and well mannered, but with gentle overtones of panic. Slowly order returned to the gallery. Bill was politely entreated not to do it again. Everyone stared at him. Laurel feels sometimes that she can't take him any-where, and he's too big to hide.

We ate well in Arles, liking particularly a small restaurant called La Gueule du Loup ('The Mouth of the Wolf', which has the idiomatic meaning of a trap). Far from being a trap, it serves inexpensive meals of good imaginative cooking by a young French chef married to an Australian girl. It is to be found in the Rue des Arènes, and will be busy on bull-fighting days.

We set out from Arles in our remaining car to prospect for a place to lay up for the rest of the winter in order to work on our boat and render it more habitable. In this way we avoided a lot of doubling back and forth at boat's speed, for the Rhône delta is as big as an English county and there are a lot of channels.

Christmas was coming up fast now, and though it is not a major French festival it is one we like to keep, and to do so in style would need a lot of work in the galley, both constructive and culinary.

We fell on our feet at the old walled city of Aigues-

Mortes, whence King Louis IX (he of the Table du Roi, and later to be the St Louis of the Blues) set sail for Jerusalem. Little wonder he got the blues, for he had two very unsuccessful crusades and died of the plague in Tunis, a sure way to be sainted. The inhabitants of Aigues-Mortes used to be known as Blue Bellies, owing to some unsavoury Midi-evil disease they were prone to.

The city created by St Louis at Aigues-Mortes is beautiful, especially at a distance. The flat land makes it possible to see the city ramparts from miles way, looking like a page from a book of hours.

A kilometre to the north, at the site of the junction of the main Canal du Rhône à Sète with the Canal du Bourgidou (now closed to navigation), we were invited, for a very reasonable sum, to berth at Le Bourgidou with the Association for the Development and the Protection of the Patrimony and Tourism of the Canal du Midi. It is a grand title, cloaking the activities of a bunch of boat enthusiasts like ourselves who devote themselves to a variety of craft, and to the restoration of the last wooden Canal du Midi barge, *Le Gaulois*. They don't receive any help from the authorities, but boat people are used to that.

'M. le Maire,' one of them (Daniel) remarked to us later, 'regards us as a pack of gypsy chicken-stealers, or worse.' We had sometimes come across the same attitude in *Fare Well*. To holiday in a yacht was considered admirably élite, but when it was found that we lived on board all the time, one could occasionally see a little cloud form in the eyes of our interlocutor, and we would be put in the pigeon-hole labelled, at best, 'harmless eccentrics' or, at worst, 'nomads, encyclopaedia salesmen and other untrustworthy people'.

Back in *Hosanna*, we had to retrace our way a couple of kilometres from Arles to the junction of the Rhône with the Petit Rhône, which in spite of its name is still a big river by British standards.

The Petit Rhône was wintry, with dead trees and branches lying part submerged or high and dry, recalling the floods of a few weeks before. The water level was low now and there were banks of sand in the meanders. Overhead, clouds were beginning to veil the brilliant sky, the weather became more misty and rainbow broughs surrounded the sun. It was cold. At the river's edge grew clumps of bamboo; these and the flood wall that protects the reedy marshes of the Camargue prevented an extended view of the countryside.

We were about halfway to the sea when we came to the next junction at St-Gilles, where a big modern lock waited to carry us out of this wild bull of a river into the peace and calm, the welcome humdrum, of a canal once more.

We entered the new lock of St-Gilles, an enormous concrete trench, a shoebox for Gargantua, 185 metres long and 12 metres broad, and were surprised to find that all this wealth of concrete and civil engineering was there to raise *Hosanna* a mere fifteen centimetres. It seemed like doing the ironing with a steamroller. Raising and lowering boats, however, is only one of its uses; another is to protect the Canal du Rhône à Sète from the Petit Rhône when it is in flood, because the former crosses the marshes of the Camargue, which are already so low-lying that it would be a disaster to flood them. There is a lot of land below sea level in this part of the world.

In the calm of a well-ordered canal once again (it will take barges of 700 tonnes), we could look about us. In the middle distance we could see a road lined with umbrella pines, and further away a mist of blue-green hills to the north of Vauvert marking the uplands of the Garrigues, and Codognan, the source of Perrier water.

Nearer the canal, old buildings with roofs of rosy Roman tiles weighted with stones marked the *vignobles*; and faded boards, roughly painted, invited us to a party that was long over – '*Dégustation et vente directe*' – in abandoned sheds whose thatched roofs had blown away since the busy days of summer, when the tourists flocked to buy the *rosé* wine of the dunes. We made our last overnight stop at the little village of Gallician, hoping against hope that the restaurant mentioned in the guide would be open. At the end of November? Of course not.

We walked round the village and the vineyards, neat fields of clipped vines, unable even to buy a drink, though the Caves Co-opérative were busy producing this year's vintage. Gallician is one of the homes of the better Costières du Gard wines, not one of the great wines of France but a topable tipple none the less. Their storage *caves* were above ground. In this marshy land where cellars would flood, they were huge circular cisterns roofed with earth, and grass and flowers are encouraged to grow there by way of insulation.

This region can produce good wine, but it needs far more work than the old *vin ordinaire*. Some of the *vignerons* go on in the old way, instead of replanting with better quality vines. The careful *vigneron* would burn his debris, but the carefree allow prunings and old uprooted vines to lie around to rot and infest the new stock. There is something of this attitude in the Camargue, a bit of the Mediterranean *dolce far niente*, combined with a suspicion of new ideas.

The last stretch of the canal was typical of the Camargue. The reeds were being cut for thatching, and were neatly stacked in piles every hundred metres, but otherwise mankind was not much in evidence. Vast expanses of flat lands extended as far as the eye could see to a huge bowl of sky.

Here and there was a clump of umbrella pines. Nearer the canal grew teasels, and tamarisk, which would be smoky-pink in spring, but was now green turning to yellow.

It was a land of birds. White egrets stood by dried puddles, or congregated in a stand of pines. Other brightly coloured birds that we did not then know, probably *guepiers*, or wasp-eaters, wheeled and hovered, mingling with flocks of gulls. Two white horses cropped the grass on the canal bank.

As we approached Aigues-Mortes, we could see for miles St Louis's lighthouse, a latticed structure on top of the Tower of Constance, the highest part of the town's great curtain walls. We thought it prudent to turn and face the next Mistral, so we had to bring *Hosanna* quite close to these walls, pierced by their massive gates and portcullises, and turn round in the *bassin d'évolution*, before proceeding back up the canal for about a kilometre, where we came to rest, facing north.

Even though at this western edge of the Camargue the Mistral is only a pale shadow of the beast that ravages the Rhône valley, the bus shelters are built with their backs to it, even if they are then at right angles to the road. The traditional *mas*, or farmhouse, stands likewise blind to the Mistral, and with small windows facing only towards the south, shuttered to keep out the heat in summer. The old thatched *cabanes* of the *gardians*, or bull herdsmen, are rounded at their northern end, to deflect the wind.

When we had snuggled into our berth among new friends we were four kilometres from the Mediterranean Sea. Near enough for our purpose. Our aim was achieved and we set to work with a will, improving our boat every day and learning about our surroundings. Many of our neighbours were former *bateliers*, working ashore now, since the decline of the *commerce*, but reluctant to leave the waterside.

One such was a sparkling lady (Martine) with teenage children, who missed her barge desperately and told us stories of her waterborne life as a child. She volunteered to drive us to Arles to look at the *bourse*, which had not been open while we were there. We accepted with pleasure, and she drove us in her little red car with the usual panache of the young French housewife, which we had until then forgotten. We hung on as the tyres screeched round sharp corners, we bounced our hats against the roof over the worst stretches of road, and listened bemused to a vivacious and rapid stream of recollections, interrupted now and again while she leaned out and yelled at another driver: 'Where are your spectacles, my little rabbit?'

We had a rewarding visit to the *bourse*, under her guidance, the more so as we were pointed in the direction of an exhibition to explain the work in progress on the Rhône. We all enjoyed it: Martine bubbled round chatting happily and swapping yarns with old *bateliers*, while we followed more slowly, absorbing facts.

Back at Le Bourgidou, we learned from the *bateliers* that the way to catch edible frogs was to 'fish' for them using a little piece of red cloth as a lure; we found the prospect of hooking our own frogs a little unappetising, however, especially after going into the cleaning and preparation process in more detail. We also learned that water hens can be eaten, but only if you skin them; we felt relieved that the pink flamingos of the Camargue, the 'Flying flowers' of Baroncelli, were not edible.

We learned, too, that the neat-looking garbage bins on the quay were NOT TO BE USED, as the long-running misunderstanding between the Association and the Mairie meant that the refuse was not collected.

Various ingenious methods of disposal were therefore in use. Martine and Mario incinerated their rubbish, while Daniel, of the *péniche Escaut*, would add his to some citizen's roadside bin when they weren't looking. We preferred to take our shameful burden to the large bins in the public places in town, though even there Laurel was caught. 'You have no right to depose the ordures in this bin, Madame!' shrieked an outraged flat-dweller from her balcony, 'It exists uniquely for the apartments!'

When visiting friends afloat in the Yacht Harbour we

would always start by nonchalantly tossing our smelly blue bag, clink, into the splendid bins provided for the tourists. Best of all solutions we found was market day, when our ordures could be sneaked into a pile of cabbage leaves, withered fruit, deflated pumpkins and broken boxes, where the municipal dustmen would sweep it all up by three in the afternoon. We of the Association took a small and innocent pleasure in outwitting M. le Maire.

At weekends we renewed our acquaintance with the Courses Camarguaises, or Bull Games.

The black bulls of the Camargue are small, wiry and quick on their feet, with long horns in the shape of a lyre. They are as different from their lumbering Spanish cousins as a racehorse is from a Clydesdale. They are driven from the ranch to the arena just as they were in the old days, four or five at a time surrounded by the *gardians* on horseback, through the streets of the town. This is called the Abrivado, and the return journey is the Bandido. The youngsters have a chance to prove their manhood in attempts to disrupt this disciplined procedure, to break the circle of horses and allow the bulls to run free, under the eyes, watchful or admiring, of their elders and girlfriends. It is expected. The *gardians* began to learn their trade in the same way when they were lads, and are proud of their skills in coping with wild bulls and wild boys.

In the Courses Camarguaises those wild boys who display particular agility and courage are selected for training, the game being to cut prizewinning tassels off the foreheads and horns of the bulls, who also learn by experience. Good bulls become star performers with well-known names – Le Toubib, Galant, Filou and Sanglier, the great star of all the stars whose tomb is marked with an obelisk. They earn large sums for their owners and are not supposed to be hurt in any way.

'What happens,' we asked, 'to those bulls who don't want to play this game?' We should have known. They are edible, and they are eaten. The Brothers Sabdes in Aigues-Mortes face each other, butcher on one side of the street and *charcutier* opposite. (Grand-mère, aged ninety, sits outside whichever establishment is most comfortable, *sol o sombra*, and totters across the street with her chair as the sun moves round). Written on their blackboards is '*Saucisson de taureau*' and '*Boeuf gardian*', which also ought to be bull meat, but in restaurants is usually cow.

'*Seiches à la rouille*' is also on the blackboard, a favourite dish of the littoral: cuttlefish in a red peppery garlic sauce. Laurel developed a milder version without the extending potato.

At the Courses Camarguaises, the white-clad young men, who walk smartly out into the sunny arena looking very like a cricket team taking the field, risk injury and sometimes death; the games may not begin until a doctor and an ambulance are present. This is a warm-weather sport, and is over by the end of October, when it finishes on a high note after the *vendange* with the *fêtes votives*, a fourteen-day holiday when the young men, fuelled with *pastis* (rows of glasses are sold at so much a metre), try even more daring raids on the Abrivados.

During the winter the Camarguais miss their summer sport, so every now and then the towns will hold an Encierro, where bulls are loosed in the streets merely for the pleasure of having them there – a street without a few bulls loose seems lonely to the Camarguais, not properly dressed – and the addicted youngsters get another bull-fix to remind them and carry them through to the spring.

We settled into a gentle country rhythm from one market day to the next, from the nuts and chrysanthemums of December through pansies and tiny artichokes to the mimosa and asparagus of early spring, with the comforting sound of boatwork all around us – the knock of hammers, the whirr of an electric drill, the rasp of saws.

As the tourist season approached, the municipality

announced, as it did every spring, the pedestrianisation of the town within the ramparts, and, as they did every spring, the *irréductibles* of the Rue Emile Jamais refused to participate and it remained the only street in the town where the sound and smell of engines intruded on the pleasant hum of footsteps and voices – and the clop and clatter of hooves, because here pedestrians include bulls and horses (and, at times of festivity, bears and camels).

As we eased in to our surroundings, we looked back on our journey and realised that we had come through many dimensions. We had come from the battlefields of Crécy and the Somme to the embarkation point for the Eighth and Ninth Crusades, from the misty woods of Corot to the brilliant sun of Van Gogh, from arctic geese and Flanders poppies to pink flamingos and salicornes, from cider and calvados to *rosé* wine and *pastis*.

We felt we had come a long way, yet we felt at home. One of the joys for us in this land is the odd resemblance of the Camargue to our own East Anglia. It is both like and unlike. The similarity is not in the vineyards and asparagus fields. Nor in the mediaeval city of Aigues-Mortes, with its perfectly preserved walls and elegant tree-lined square with the statue of St Louis, where on good days in winter we sit in the sun and drink coffee, watching the world go by (you can't do that in Beccles). Nor in the weather, which, though it can be cold in winter, seldom freezes and is often warm and sunny. Nor in the torrid summer heat, when the shade of trees is sought by all but the tourists.

What is truly reminiscent is the flat open countryside, the marshes and salt flats and the huge brilliant sky; the Mistral, the big brother of the north-east wind of Norfolk, blowing sand in through the letter-box; the birds, and boats on the rivers and canals; and the people, who are disgracefully obstinate and insular and proud of it.

We appreciate a land where tourism is comparatively new and the social life of the community does not yet

revolve around the needs of the visitors, but is centred firmly on the pleasures of the inhabitants, for generations hard-working, hard-riding, hard-living agrarian cowboys, salt-miners and shepherds – not Provençale, not Occitan, not Catalan, but Camarguais, a race apart from the rest of France. Before we move on, we shall enjoy this land.

Appendix 1

A taste of France

'Oh! de bon Dieu! Li sartanado einormo
de sang de biou, li tian de tripo grasso,
Lou catigot, Li carbounado e chouio,
E li troucho farcido eme do berlo,
s'engloutisson i buerbo pansarudo.'
....

'Oh! dear God! the enormous
fricassées of black pudding, the
tians of tripe, the fish stews, the
carbonnades, the grills, the omelettes
filled with wild celery, drowning in
spacious bellies.'

Only one of these recipes is difficult. Some are decidedly
basse-cuisine for benighted *bateliers* in the depths of winter.
Some are an echo of summer, for lunch on a warm day. All
recall for us the sound of birds and the ripple of water, or
the whiff of smoke from our wood stove.

Seafood *ficelles* (Picardie)

These pancakes, similar to those we ate at the Hôtel du Port
at St-Valery, are stuffed with seafood, coated with a sauce
and bubbled under the grill with cheese 'strings', hence the
name *ficelles*. They are quite filling.

Make some thin pancakes; a standard batter of 4 oz flour,
1 egg, and ½ pint of milk will make about 10. Stack them,
interleaved with foil or kitchen paper, and put aside.

For the filling:

> ½ lb firm white fish (cod, haddock or monkfish)
> 1 lb mussels
> 2 oz cooked, shelled shrimps

Cook the mussels; see recipe below on pp. 209–10. Poach the fish. (The monkfish, or angler, is the one with the ugly mug, so much so that it is always headless on English fishstalls in order to cause no nightmares, and ranged belly-up on continental stalls, where they know you will need the head for stock. It is a large head, and its teeth tend to bite your fingers as you try to fold it into a saucepan. An excellent fish, the best for the purpose, it is now expensive and hard to find in England, since the French (who call it *lotte* or *baudroie*) will happily pay the equivalent of £13 a kilo for it.)

A microwave is useful for poaching as it is easier to avoid overcooking the fish, and you need not add liquid. The fish broth that results can go into the sauce. If the mussel juices are not too salty, they can also be used. Shell the mussels, remove and discard any skin and bones from the fish, and put mussels, fish and shrimps aside to keep warm but do not allow them to dry up.

For the sauce:

> 1 oz butter   approx ½ pint milk
> 1 oz flour   a little fish broth

Melt the butter, add the flour, cook and stir for a moment. Add the broth, allow to boil and stir. Add enough milk to make a thick sauce and cook for a few minutes, stirring. Season. Use enough of this sauce to glue together the flaked fish, mussels and shrimps, with a little chopped parsley. Put some of this mixture on each pancake, roll them up and put them in an ovenproof dish.

Now add more liquid to the remaining sauce, until it is of

a mind to pour itself over the pancakes (check the seasoning), then sprinkle with grated cheese – Cheddar will do
nicely if you have no Gruyère, but might not look so
stringy. Brown under the grill, and serve to four people,
two of them hungry.

Mussels and chips (Picardie)

This is equivalent to fish and chips in England, and is
available at all coastal resorts of the Nord-Pas de Calais and
Boulogne area, and quite a way inland into Picardie, as far
as Ham and Péronne.

Mussels are not an expensive food, even in France,
where they cost in the fish shop twice as much as they do in
England. In a restaurant they make a cheap lunch. They
contain trace elements that are very good for you, and no
cholesterol.

You don't have to eat the chips, but if you want real
pommes frites, cut the potatoes fine and deep-fry them twice.

Put two pounds of mussels into cold water, and get
someone who loves them to scrape and beard them. Beard-
ing should be done with care, as this bit of old brown string
is the mussel's filtration system and contains the rejected
impurities. You, too, should reject them. Scraping is partly
a matter of aesthetics, and partly to get the sand out; if
you don't need the shells in the soup, you can skimp on
this. You can always keep a few highly polished shells to
sprinkle around for effect, and wash them afterwards for
re-use. A bit of a scrub to remove the sand, and a rinse or
two, is a good idea. If an open shell does not respond to a
gentle squeeze by closing, put it aside. When you have
finished all the others give it a last try, and discard if
moribund. Awful as it sounds, mussels should be alive
when cooked; only thus do you get the plump delicious
little morsel as it should be. A mussel that succumbed
before cooking will look shrunken, wrinkled and small in

its shell, which gives you another chance to reject it before eating it.

Put the mussels in a large pan, with some chopped parsley and onion if you like. No water is needed, but if you want the liquor for soup, throw in a small glass of dry white wine. Cover the pan, and put it on the heat, and when steam blows from under the lid count sixty, shaking the pan. This is all the cooking they require.

Take the pan off the heat and remove all mussels that have opened, giving the unopened ones a little longer to cook. Shell them, and cool. The juice will amost certainly be very salty and probably sandy, but can with care be used for soup (don't use much), or with onions, white wine and parsley for *moules marinière*. Try the bigger mussels just as they are with a good mayonnaise and French bread. Or pick out the largest, leaving the bottom shell on, and stuff them like snails with:

Bourguignon butter

1 oz butter
1 tbsp finely chopped shallot or onion
1 clove crushed garlic
1 tsp chopped parsley
1 tbsp of breadcrumbs

Fry this all together in a small pan for a minute or two, then use it to stuff the mussels, and pop them under the grill to warm.

Freezing does not seem to suit mussels: the texture suffers and they end up like flannel rather than pleasantly bouncy. Fortunately, once cooked and cooled, they will keep in the fridge for several days, well covered to keep them moist.

Chocolate mousse

The three puds most commonly available on the sort of menu we usually eat in France are *crème caramel*, ice cream or chocolate mousse. All of them can be delicious or awful, according to whether they are properly made or shaken out of a packet of chemicals like Artemis springing fully armed from the thigh of Jove. It seems to us that the little extra time and money it takes at the family level to use good ingredients is worth while. So, take:

> 4 oz bitter chocolate
> 4 eggs, separated
> 2 tsps brandy and two tsps Cointreau (or 4 tsps strong coffee)

Melt the chocolate gently with the alcohol or coffee in a bowl over boiling water, stirring carefully till it becomes glossy. Do not overheat or it will become irreparably grainy. Remove from the heat and stir in the four egg yolks. Whisk the whites with a pinch of salt until stiff, and fold gently into the cooled chocolate until no white specks remain. Pour into six ramekins and chill. No sugar is added, as we like our chocolate dark and bitter.

Amiens duck *pâté*

'Put in a recipe for the duck *pâté* you had at Amiens,' said our editor. 'Why not?' I said. Rash fool. This is the difficult one. I remembered doing a splendid duck *pâté* for Christmas many years ago, so I sailed into the preparation of this one without qualms. I had either forgotten the time and trouble it took, or I was a better cook in those days. On this occasion I chose the oldest of my recipes as being the most authentic (apart from the original Amiens recipe which wrapped a duck in pastry, bones and all, a bit mediaeval for the 1990s.

1 duckling, about 3½ lbs
6 oz lean pork or veal
4 oz chicken livers
2 oz ham or bacon
1 small onion
4 oz mushrooms
1 tbsp brandy
1 egg
12 oz shortcrust pastry
4 oz jellied stock

Ask the butcher to bone out the duck, leaving the skin whole. Mince or chop finely the pork or veal, the chicken livers, the duck's liver, the ham or bacon, and the onion and mushrooms. Moisten this forcemeat with the brandy and beaten egg. Season fairly strongly. Stuff the duck with this, re-forming the shape and sewing up the orifices.

Roll out the pastry (8 oz plain flour, 4 oz butter, water to mix) and line a greased 1-lb loaf tin, or a suitably sized terrine, with two-thirds of it. Place the duck therein, and use the rest of the pastry to form a lid over the top, sealing the edges well together. Make a hole in this lid. Brush the pastry with a little beaten egg. Set the tin in a pan of water and bake in a moderate oven (350 degrees F, 180 C) for 1½ to 2 hours. Pour the warm jelly in through the hole to fill any remaining gaps, and allow to set. This is the *best* way to do it.

Sounds so easy, does it not. I was doing fine until I found that my boned duckling was not obeying the rules. It seemed to be wearing its Daddy's overcoat, which was much too big, and in order to re-form the bird's shape I had to take huge tucks and gussets in its skin. Sewing it up was not as easy as it sounds either – it reminded me of wound-stitching in the Ship Captains' Medical Course, where we practised on foam rubber, since real skin is elastic, resistant and extraordinarily hard to stitch. So was this duck.

None of the recipes I consulted made any mention of tins or terrines, but enjoined one merely to cut two ovals of pastry and encase the duck therein. This I did, and watched aghast, having sealed my pastry edges well together as instructed, while my *chef d'oeuvre* (one could not call it an *oeuvre du chef*) sighed, slowly spread itself, and settled into a one-inch high pancake the size of a dinner plate.

Drastic measures had to be taken. I used my common sense and eased the whole thing into a greased loaf tin, into which it subsided with gentle protests in the form of sucking noises and bubbles blown through the hole.

I would not have dared to give the recipe here if I could not assure you that it ended up a success.

Tarte aux fruits

Every *pâtissier* in France makes these deliciously rich and quite expensive desserts, so they do not appear on the menu in cheap restaurants. They are, therefore, well worth making at home.

For the sweet short crust:

    8 oz plain flour
    1 tbsp caster sugar
    5 oz butter
    1 egg
    A few tsps water

Make up into a pastry dough and chill for at least an hour. Then roll out and line a greased 7-inch flan dish, prick the base with a fork, and bake blind (line the tart with grease-proof paper and fill with breadcrusts or haricot beans to keep the base flat) for about 15 minutes in a hot oven.

For the crème pâtissière:

> 2 egg yolks
> 2 tbsps cornflour
> 1 vanilla pod
> 2 oz caster sugar
> 8 fl oz milk

Beat the yolks with the sugar until thick and creamy. Mix the cornflour with 4 tbsps of the milk and incorporate with the egg mixture. Boil the rest of the milk with the vanilla pod, remove from heat and infuse for 10 minutes. Remove the pod, add the milk to the egg mixture, and heat gently to boiling point. Allow to boil for 2 minutes, stirring all the time. Remove from heat and beat rapidly; it should now be quite thick. Cool, and spread over the base of your *tarte*.

Now you can let your imagination rip, and make a *tarte* of your own *maison*. Because of the custard base which isolates the juice from the pastry, you can use fresh fruit such as strawberries, raspberries, kiwi, bananas and melon, or all of them in a colourful medley. You can use drained tinned fruit, such as apricots. You can use stewed pears, plums or gooseberries. Lay the chosen fruit in a decorative pattern on the custard base, and if you want it to look really professional cover the fruit with a jelly glaze. (You can cheat with a packet, or use strained apricot or redcurrant jam.)

There is only one rule: this *tarte* should be eaten on the day it is made; no difficulty should be experienced in persuading someone to eat the last crumb.

Potage Crécy (Picardie)

This carrot *purée* soup is claimed by two Crécys, one in the Somme (Picardie) and one in Seine-et-Marne, nearer to Paris. I felt it belonged to Picardie, which grows such splendid carrots, and that is where I made it, near Abbeville.

4 oz sliced carrots
2 tbsps chopped onion
1½ pints stock, milk and water, in any proportion
3 oz rice
2 oz cream

Fry the vegetables gently in butter, without letting them brown. When the onions are transparent, add all the vegetables to the stock and simmer for about 20 minutes or until the carrots are tender. Add the rice, cover the pan and cook gently for a further 20 minutes. Then put the soup into the blender or through a sieve. Reheat with the cream, correct the seasoning and serve with *croûtons*. This serves four, so I put half of it in the fridge for the next day's Sudden Soup, which can be beefed(?) up with slices of smoked boiling sausage or chunks of ham.

Quiche Irène (Aisne)

Let us thankfully forget the fast-food quiche, soggy, overcooked, thin and tasteless. Real men may not eat quiche, but men eat real quiche.

6 oz short crust
¼ pint milk and ¼ pint fresh thick cream, (or ½ pint cream if you like)
3 newlaid eggs, 1 tsp salt
1–2 oz smoked bacon or *petit salé* (salt pork belly)
Grated Gruyère or Cheddar

Roll out the short crust (bought if you're rushed, but it's almost quicker to make some than defrost it; see quantities on p. 212), and line an 8-inch greased flan tin or round oven dish with it. Separate one of the eggs, and use some of the white to paint over the pastry base; this helps to stop it becoming soggy. Chop the bacon and lay on the base. Beat the remaining three yolks and two whites with the milk and cream, and season with salt and a little pepper. You can add grated cheese, either to the mixture or sprinkled on top.

As this is basically a custard, it needs gentle treatment (40 minutes in a moderate oven) if it is not to curdle. If you notice the filling belch like a hot mudhole, turn the heat down. Serve warm, and it is better reheated in the oven rather than the microwave. Serves four to six.

I cooked one of these once in the Bay of Biscay when there was the usual gale blowing. We had a vegetarian on board, and I took care to place only mushrooms in his section of quiche, before placing it carefully in the oven. Unfortunately, before the eggs had set we rolled heavily, and several renegade chunks of bacon invaded his portion. He was annoyed. And I was annoyed that he was annoyed. Cooking in a gale is pretty hard work, but we do not have the rule that any person who complains cooks the next meal as food is too important, especially in bad weather.

There is a well-known story of the boat where this rule obtained. The most recent complainer cooked as badly as he could in the hope that someone else would soon take his place, but everyone chose their words with care until at one meal a sock appeared in the stew. The luckless recipient said without thinking, 'What's this?'

'Any complaints?' said the cook, bright-eyed.

'No, no . . . it's . . . umm . . . perfectly cooked.'

Irène's *pot-au-feu* (Aisne)

If you have a wood stove, this can simmer gently on the stove top all day.

> 1½ lbs shin or brisket
> A chunk of marrowbone
> 2 carrots
> 1 small turnip
> 2 medium onions
> The white parts of 2 leeks

2 sticks celery
bouquet garni
2 cloves of garlic
3 pints stock
Quarter head cabbage

Put the pieces of meat and bones in the stock and bring to the boil. Cut the vegetables into even pieces, and add them with the garlic, if liked, and the *bouquet garni*. Skim off the froth and simmer for four hours, adding the cabbage for the last half hour. Serve the broth separately, adding rice or vermicelli to it if you wish, and eat the meat and vegetables with French bread, gherkins and mustard.

Snails, real and mock

If you find and gather your own snails (but not, we hope, in your shoes, as we did at Rachecourt-sur-Marne), do not pick them off ivy plants, as these give them an acid flavour. Look for the *petit gris*, which is grey with white flecks, or the Burgundy snail, pale brown with little stripes. A third edible variety has stripes of chestnut and white.

Put the little dears in a flower-pot to fast, covering the top with something to prevent escape, but allow ventilation. Leave them for several days, ignoring their cries of hunger. Then wash them in plenty of water, and purge them by sprinkling with a spoonful of flour and a small glass of vinegar. Leave overnight. Wash them again, blanch them in boiling water, drain and cool. Take them out of their shells and remove the little black ends. Cook them in an uncovered pan, with some wine and water, chopped onion and parsley, for about three hours. Boil the shells with a streak of vinegar in the water until they are clean. Replace the snails in their shells.

Or save yourself a lot of work by buying them in a tin.

You can also use mock snails made of a firm piece of

mushroom. Pack the shells with Bourgignon butter (see above under mussels, omitting the breadcrumbs), and warm in the oven. You can buy disposable snail plates, which hold the shells upright so that the sauce does not empty out. I speak as one who has attempted to prop them up in various other ingenious but completely unsuccessful ways.

Water feast

Dreamed up in the tunnel of Balèsmes, we have eaten the elements of this feast at various times and places, but never, we have to confess, all together.

Watercress soup

1 bunch watercress
1¾ pints stock or water
4 oz potato, chopped small
1 leek, chopped small
1 tbsp chopped onion
Béchamel sauce (made with ½ oz each of butter and flour, and ½ pint of milk)

Cook the potato, leek and onion in the stock or water for 15 minutes or until soft. Wash and pick over the watercress, blanch it with boiling water, drain and chop finely. If you have a blender, whizz together the stock, Béchamel sauce and watercress. (Do not be tempted to whizz the cress without first chopping it, as the result is similar to a rope round your screw on the waterway – the stems jam round the blades.) In the absence of a blender, *purée* the vegetables with a fork or potato-masher.

Reheat the soup, add cream if you like, adjust the seasoning, and serve to four persons strewn with reserved sprigs of watercress or chopped chervil. (The soup, not the per-

sons). You will observe that the cress is scarcely cooked at
all, which gives this soup a pleasantly fresh taste.

Trout
There is nothing better than a fresh trout fried in butter.
Except, some might say, two fresh trout fried in butter.

Duck
Among the hundred ways of cooking this appealing bird,
the simpler ones seem to produce results as good as the
complicated. I like to roast it with a baste of honey and soya
sauce, and hope that there will be some duck left over to eat
cold.

Waterway vegetables
WATERLILY POTATOES are made from the roundest baking
potatoes you can find. Cut a deep cross in the top of each
one, then bisect the cross with four more cuts, making a star
which will press up to form a waterlily when the potato is
baked.

WATER CHESTNUTS: usually tinned. Drain well, slice in two
horizontally to form coin-sized 'chips', and *sauter* (fry in
butter) as if they were potatoes.

SNOW PEAS. (*Mangetouts*) These keep their fresh flavour and
emerald colour best if cooked in the microwave oven.
Sprinkle with a teaspoon of water, dot with a hazelnut of
butter, cover and cook for a couple of minutes to keep them
slightly crisp, poke them with a knife point to determine the
amount of resistance, and do them a little more if you
favour them less crunchy.

Floating Island pudding
Floating Island pudding (*Ile flottante*) is one of the desserts
on offer in French restaurants if you go up a notch in price,

along with the *tarte maison*, raspberry *bavarois*, and, if you are really unlucky, frozen Black Forest *gâteau*.

Interestingly, Floating Island pudding has changed. It used to be a decorated sponge cake floating on a sea of egg custard. Perhaps it is the trend to lighter desserts that has produced the feathery meringue islands of today.

4 eggs, separated
1 pint milk
3 oz caster sugar
1 vanilla pod

Beat the yolks of the eggs with the sugar until thick and creamy. Heat the milk with the vanilla pod. When it boils, remove the pod and pour the milk on to the egg mixture, stirring constantly. Put this custard back on the stove and heat very gently, stirring all the while, until the mixture coats the back of the spoon. Put into a large bowl or four small ones and set aside to chill.

Whip the whites with a pinch of salt and a further two tablespoons of caster sugar until they are stiff. Now we play boats. On to the surface of a large pan of simmering water slide large spoonfuls of the meringue. Treat them like a fleet review, and do not let them touch each other. Cook them for about 6 minutes, turning them over at half time. Remove with a slotted spoon and drain on kitchen paper. Float them gently on to the custard just before serving.

Watermelon and feet-in-the-water blackberries

½ watermelon
¾ lb blackberries
sugar
lemon juice

Cut the watermelon into one-inch cubes, discarding the black seeds, and place in a bowl. Add the blackberries. Sprinkle the fruit with a little lemon juice and sugar. Chill.

Bread

My high-protein bread recipes for transatlantic crossings would be out of place in this book, but here is Lost Bread.

Pain perdu

This is a recipe known worldwide under different names: 'Torrijas' in Spain, 'Poor knights of Windsor' in England, 'Rich knights' in Finland, and 'French toast' in America, which would amuse the French.

Take squares of crustless bread (*brioche* is used in France, very rich and delicate) and soak them in ¼ pint of milk into which you have beaten an egg. When the slices have been well dunked, fry them in butter. At this point the whole thing is sometimes ruined by rolling the slices in icing sugar, which makes them very floppy. Turn them over in *granulated* sugar and cinnamon, for a crunchy contrast to the egg. The alternative is to spread them with good apricot jam.

Boeuf en daube

Every region in France has a version of this long-cooking winter dish, to simmer on top of the wood stove.

> *For the marinade:*
> ½ pint red wine
> ½ pint water
> 1 sliced carrot
> 1 sliced onion
> A *bouquet garni*

Into this put 1 lb stewing beef cut into chunks, and leave it there for a couple of hours. Then take out the meat, drain it and roll it in seasoned flour. Put it in a stewpot with a heavy lid, (the original *daubière* had a dished lid that was filled with hot cinders).

Bring the marinade and its contents to the boil, and cook

a crushed garlic clove in it for 10 minutes. Add to your stewpot 2 oz bacon, diced. Slice two more carrots and another onion, and fry lightly. Put these and a *bouquet garni* in the pot. Strain the marinade over the meat, cover tightly and simmer gently for four hours.

Pork chops *Avesnois* (Nord)

This is a quick and easy way of doing pork chops, which are good everyday food and sometimes need some variation. Grill the chops. When almost done, spread on them a paste made of grated Cheddar or Gruyère (3 oz cheese will do four chops), and equal parts of cream and French mustard, adding just enough of this goo to the cheese to make a stiff paste that will not slip off the chops when spread upon them and browned under the grill.

That's all. Delicious.

Salmon steaks with sorrel sauce (Burgundy)

We had these for Bill's birthday supper at Port d'Arciat.

> 4 salmon steaks
> 4 fl oz dry white wine
> 4 fl oz fish stock
> ½ pint double cream
> 2 handfuls fresh sorrel

I never got salmon steaks right until I had a microwave oven. Now I can watch them turning to just the right pink. Otherwise, poach them in a little fish stock. To make the sauce, heat the wine and stock in an enamel pan, cooking it down to about 4 tablespoonsful. This requires careful watching, so avoid interruptions. Add the cream and continue to reduce until the sauce is a good consistency. Season. Tear the sorrel into little pieces and stir it in, with a knob of butter, just before serving. A final squeeze of lemon juice can be added. This sauce is just as good with salmon trout.

Soupe de poissons

This fish soup is sieved, simple and savoury.

1¾ pints water
1¼ lbs soup fish (*poissons des roches*)
1 leek
1 small onion
2 crushed cloves of garlic
A small piece of celery
A frond of fennel
A bay leaf
A *bouquet garni*
A pinch or two of saffron
A handful of parsley, pounded
2 tbsps tomato paste

Although the soup is simple to make, there is an art in choosing the fish to go in it. The colourful pile of mixed small fish known as *poissons du roche*, or *du golfe* – red *rascasse*, silverblue sardines, the brown stripes of *vivre* (weever), the rainbow colours of the wrasse family – does not appear in the British fish shop. Try any small bony white fish, and include gurnard, wrasse, scorpion fish (*rascasse*) and a bit of eel if you can get them.

Cut up the vegetables and fry gently in a little olive oil. Stir in the tomato paste. Add the water and the aromatics: the crushed garlic, bay leaf, fennel, *bouquet garni* and the pounded parsley.

Clean the fish but leave the heads on. Throw them into the water and simmer for 15 minutes, then sieve the soup, preferably through a piece of muslin, to catch the tiny bones. Return to the pan, season quite strongly with salt and pepper, and at this stage add the saffron. Reheat, and serve with toasted slices of French bread, *rouille* (see recipe below) and grated cheese.

Sardine *bouillabaisse* (Provence)

> 1½ lbs fresh sardines
> 1½ lbs firm potatoes
> 2 pints water
> 3 tbsps olive oil
> 2 tomatoes, or 2 tbsps tomato paste
> 1 leek, chopped
> 1 onion, chopped
> Fennel
> Orange peel
> Bay leaf
> 2 cloves garlic
> Saffron
> sea salt
> pepper

Make a broth by adding the tomato, onion, leek and olive oil to the water. Crush and add the garlic; the gentle rose-pink cloves of Provence would be nice. Flavour with a frond or two of fennel, an inch of orange peel, the bay leaf, and salt and pepper. Bring to the boil, add the potatoes cut in thick slices and cook for 15 minutes, or until the potatoes are nearly done. Do not allow them to break up.

The form in which you add the sardines is up to you. If you are fearless, behead and clean them, and tip them in to cook for the last 5 minutes. Or take the backbones out and do likewise. For really easy eating, fry them separately, then pass them through a sieve and add the *purée* to the soup. At the very end, stir in a pinch of saffron. Serve with slices of French bread which have been rusked in the oven, and spread with *rouille* if you have it. Astonishingly good.

Scotch pancakes

I have seen these in a French cookery book, called *crêpes Ecossaises*, so perhaps it is not altogether cheating to include them. They are synonymous with winter outside and warmth within, the need for something to comfort the flagging spirit.

> 4 oz flour
> ¼ pint milk
> 1 egg
> ½ tsp cream of tartar
> ¼ tsp bicarbonate of soda

Beat the ingredients together to form a smooth batter. Drop in spoonsful on a greased griddle or flat-based frying pan. When the tops lose their glossy appearance, flip the pancakes over and gild the other side. Stack them with kitchen paper between the layers to prevent drying out, and keep warm while you cook the rest. Serve warm with butter and golden syrup. Your fingers will be sticky, but the world suddenly looks a better place.

Seiches or *poulpes à la rouille* (Provence)

Now we are really in Provence. This is an appetiser, redolent of hot sun and salt sea. *Seiches* are cuttlefish. *Poulpes* are baby octopus. You can use ready-made frozen rings of squid if you like. You will also need a jar of *aioli*, and another of *rouille*. *Aioli* is mayonnaise with crushed garlic added, known as 'Provençale butter'. *Rouille* is a garlicky sauce, fiery with mustard and hot peppers, which give it a beautiful persimmon colour.

Rouille
A jar of *rouille*, that excellent accompaniment to Mediterranean fish soups, should be in the family's baggage along

with the duty-frees at the end of your holiday in the Midi. Failing that, it is made as follows.

Pound together in a mortar 2 cloves of garlic (more if you like) with two small red chillis (from which you have removed the seeds), until well combined. Add a piece of bread, dipped in fish broth and squeezed almost dry, and pound this into the mixture. Then add to it about 2 tbsps olive oil, a little at a time. Finally add a little fish broth until your *rouille* is the consistency of mayonnaise.

> *For the* seiches:
> 8 oz cuttlefish (or calamar, octopus or squid)
> 1 onion
> 1 clove of garlic
> two tomatoes, peeled and seeded
> 1 tbsp *aioli*
> 1 tbsp *rouille*

Prepare and clean the fish if necessary, removing the transparent stiffener of the squid, the bone of the cuttlefish, or the beak of the baby octopus. (Now you know why fishing boats are so often called 'Rien sans peine' – Nothing without trouble.) Cut the flesh into pieces about an inch square, and fry gently in olive oil. When you have evaporated all water, add a chopped onion, a crushed clove of garlic, a pinch of thyme, a crushed bay leaf, the two chopped tomatoes (or 1 tbsp of tomato paste), and cook a little longer.

Moisten with a little wine and water if it seems to be getting too dry. Mix a tablespoon of *aioli* with an equal amount of *rouille*, and stir into the pan. The chunks should neither be too dry, nor swamped with sauce. Add some more if needed.

This makes a strongly flavoured appetiser, which goes a long way. It is a lovely red-gold, and a spoonful nestles nicely into a lettuce leaf. Have a *pastis* with it and recall the hot sun and sea salt of the Mediterranean.

Appendix 2

HOSANNA
General arrangement

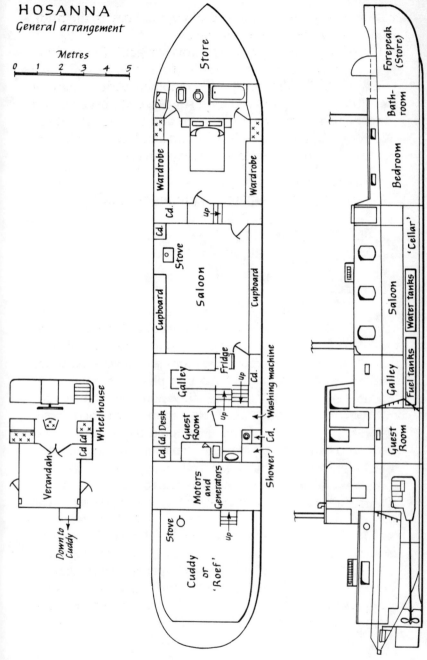

Appendix 3

Some information for those contemplating a canal cruise

There are not many useful guide books, and most of those published in England suffer from trying to cover too much ground in one volume, and thus end up by not covering it well enough. Those published in France tend to be more detailed, but still do not manage to keep up-to-date. Here, however, is a list of those books that most helped us to enjoy our voyage, though it is by no means a complete bibliography.

Practical guides

Cruising French Waterways by Hugh McKnight (Stanford Maritime, 1984). This is a guide book of background information in pleasant narrative form, but already sadly out of date and very much summer-time oriented. The author clearly knows some parts better than others, and has the grace (rare among the writers of guides) to admit that some of his research was carried out with an outboard-powered inflatable. Given the problem of trying to cover the whole of France, this guide is probably the best, especially while travelling.

Inland Waterways of France by David Edwards-May (Imray Laurie, Norie and Wilson, 5th Edn, 1984). This is a revision of a classic work by E. E. Benest. It is mainly a rather bald collection of figures, mostly distances. It could be useful at a preliminary planning stage, but we did not find it much use while on passage.

Cartes Guides de Navigation Fluviale (Editions Cartographiques Maritimes) and *Guides Vagnon de Navigation Fluviale* (Guides et Cartes Vagnon, Les Éditions du Plaisancier) are both published in France and are available in French, English and German, though in the older editions the English is, to say the least, quaint. In the later editions there is some input from the hiring agencies, which is not always what the private-boat-owner needs. The guide parts are a little bald, and owe something to municipal tourist brochures. The maps are well drawn so far as they go, but are in strip form, shy of surrounding topography, and some contours or spot heights would be welcome. All become out-of-date very quickly, and are very inaccurate on the vital subjects of depths, moorings and bridge clearances, and (especially in the Canal du Midi) of the shape of those bridge clearances. However, one or other of these guides is more or less essential for the full enjoyment of a cruise. Obtainable from Stanford's Map Shop in Long Acre, London, or more cheaply in France if you can find them.

General books

France – the Quiet Way by John Liley (Stanford Maritime, 1975). An atmospheric book with nice illustrations of the waterways.

A whole series of books about canal travel in France by Roger Pilkington, published by Macmillan in the sixties, also give a good feel of the subject.

Hiring a boat

Those contemplating hiring a boat for a holiday will find there are innumerable agencies advertising in the French magazine *Fluviale**. Alternatively, Hoseason's of Oulton Broad, Lowestoft, is a British agency offering good boats for hire in France, and there are of course many French hire-boat companies.

There are also several operators of so-called hotel-barges, where a large *péniche* has been converted into a moving hotel with cabin accommodation for guests, and all naviga-tion is done by a professional crew. These vary in quality from just above the squalid to the very luxurious, the latter sometimes being accompanied by mini-buses to take the clients sight-seeing. We have ascertained that at neither end of the scale is this business as profitable as it appears, with the consequence that operators go out of business regularly, often being replaced with a new team of en-thusiasts about to learn the economics of operating a boat that is licensed to carry passengers. Because of this turn-over we do not feel able to recommend a particular oper-ation, and feel it would be better to consult the pages of *Fluviale*, or the British magazines that cater for inland water enthusiasts: *Motor Boat and Yachting* and *Waterways World*. Advertisements are also to be found occasionally in the holiday pages of the 'quality' Sunday newspapers.

Buying a boat

If you are considering buying a barge we would re-commend the use of an agency. Barges for conversion are not very expensive, and though the commission sounds excessive in percentage terms (usually 10 per cent) it is a fair reflection of the work involved in this case.

We bought *Hosanna* through the agency of Mr Nickel Rijk, of Friesland Boating, de Tille 5, 8723 ER Koudum, Holland, Telephone 05142 2607. He speaks excellent

---

*Obtainable by post from Editions de l'Ecluse, 64 Rue J. J. Rousseau, 21000 Dijon, France.

English. Another who is known to us by his Christian name only is Gerard, of the agency H$_2$O, Port de Plaisance, 21170 St-Jean-de-Losne, France. Telephone: 80 39 23 00 or fax: 80 29 04 67. He also speaks excellent English and German.

There are considerable numbers of French canal barges for sale cheaply, and these are often advertised in *Fluviale*. Though the prospect of managing thirty-eight metres of vessel may seem daunting, French *péniches* are constructed as a sort of shoe-box of standard section (as opposed to the Dutch, which tend to have a gracefully curving sheerline), and thus lend themselves to easy and inexpensive shortening. In many cases anyone wanting a boat of, say, 20 metres would find it a better proposition to cut the middle out of a French *péniche* than to start from scratch. Again, from our own experience, we found Bure Marine of Cobholm, Great Yarmouth, of great assistance to us in the practical work of converting *Hosanna* at reasonable cost.

Whether you hire, borrow or buy, we wish you luck and *bon voyage!*

Index

Index